Riddle of the Rhine

Chemical Strategy in Peace and War

Also from Westphalia Press

westphaliapress.org

Riddle of the Rhine

Chemical Strategy in Peace and War

by Victor Lefebure

WESTPHALIA PRESS
An imprint of Policy Studies Organization

Riddle of the Rhine: Chemical Strategy in Peace and War
All Rights Reserved © 2013 by Policy Studies Organization

Westphalia Press
An imprint of Policy Studies Organization
1527 New Hampshire Ave., NW
Washington, D.C. 20036
dgutierrezs@ipsonet.org

ISBN-13: 978-1935907558
ISBN-10: 1935907557

Cover design by Taillefer Long at Illuminated Stories:
www.illuminatedstories.com

Updated material and comments on this edition
can be found at the Westphalia Press website:
www.westphaliapress.org

THE RIDDLE OF THE RHINE

THE LIVENS PROJECTOR—I.

A completed battery of projectors in the foreground, with a battery on the left rear, half dug in. Suitably camouflaged with brush, the batteries are not observable by aircraft, and, being in "No-Man's-Land," neither party can detect them by day.

Frontispiece

THE RIDDLE OF THE RHINE

CHEMICAL STRATEGY IN PEACE AND WAR

An account of the critical struggle for power and for the decisive war initiative. The campaign fostered by the great Rhine factories, and the pressing problems which they represent. A matter of pre-eminent public interest concerning the sincerity of disarmament, the future of warfare, and the stability of peace.

BY

VICTOR LEFEBURE

Officer of the Order of the British Empire (Mil.)
Chevalier de la Legion d'Honneur, Officer of the Crown of Italy
Fellow of the Chemical Society, etc.

WITH A PREFACE BY

MARSHAL FOCH

AND AN INTRODUCTION BY

FIELD-MARSHAL SIR HENRY WILSON, BART.
Chief of the Imperial General Staff

THE CHEMICAL FOUNDATION, INC.
85 BEAVER STREET
NEW YORK CITY

PREFACE

My motives in writing this book are sufficiently explained in the first chapter. The silence surrounding the true facts of the chemical campaign, the tardy realisation of the real forces behind it in Germany, and our failure to grasp the significance of the matter in the Treaty, all pointed to the need for an early statement. More recently, this need has been emphasised by inaccurate public utterances on the matter, and by its vital importance for the full and fair treatment of certain legislative measures before Allied countries.

A unique experience of chemical warfare in all its aspects, first with a combatant gas unit on the British front in France, then as Liaison Officer with France and other Allies on all Chemical Warfare and allied questions, has afforded me an exceptionally complete survey of the subject. Later post-armistice experience in Paris, and the occupied territories, assisting Lord Moulton on various chemical questions in connection with the Treaty, and surveying the great chemical munition factories of the Rhine, has provided a central view of the whole matter which can have been the privilege and opportunity of very few.

Further, my association with the dye industry, since commencing this book, leaves me with a deep conviction of the critical importance for disarmament, of a world redistribution of organic chemical production. It is inevitable that such a step should benefit the growing organic chemical industries of

5

Preface

countries other than Germany, but this issue need
not be shirked. The importance of the matter is so
vital that it eclipses all reproach that the disarmament argument for the maintenance of the dye
industry is used on selfish grounds. Such reproach
cannot, in fairness, be heard unless it destroys the
case which we have established. We are faced
with the following alternatives. Safety demands
strong organic chemical industries or cumbersome
and burdensome chemical warfare establishments.
The stability of future peace depends upon the
former, and the extent to which we must establish,
or can abandon, the latter depends entirely on the
activity and success of those whose special duty it is
to organise against war.

A recent visit to America revealed the considerable publicity and public interest surrounding chemical warfare, strengthening my conviction that the
facts, now noised abroad, should be presented in
their proper setting. They are supremely significant
at the present time and for the future, hence the
chapters which follow.

V. LEFEBURE.

HAMPSTEAD, *October 12, 1920.*

6

PREFACE BY FIELD MARSHAL FOCH

In 1918, chemical warfare had developed considerably in our army. Before 1914 Germany possessed chemical factories which permitted her to manufacture in great quantities chemicals used at the front, and to develop on a large scale this new form of fighting.

The Allies, to retaliate, had to experiment and organise important centres for production. Only in this way, though starting late, were they able to put themselves in a position to supply the growing necessities of their armies.

To-day, the ability for aviation to carry increasing weight furnishes a new method for abundantly spreading poison gases with the aid of stronger and stronger bombs, and to reach armies, the centres of population in the rear, or to render regions uninhabitable.   -

Chemical warfare is therefore in a condition to produce more formidable results over more extended areas.

It is incontestable on the other hand that this growth will find an easy realisation in one country, Germany, addicted in times of peace, to wholesale manufacture of chemical products, which a simple modification in reactions can transform into war products.

This country, deprived, partially at least, of its former methods of fighting, and its numerous forces of specially trained soldiers, regularly organised and

7

strongly armed, will be more drawn toward the new systems of attack—that of chemical warfare.

Chemical warfare must therefore enter into our future provisions and preparations, if we do not wish to experience some terrible surprises.

The work of Major Lefebure gives an exact idea of the possibilities he finds to-day in Germany, and through them the dangers with which she threatens us. In this form it constitutes a warning, and information of the highest order, for the minds who remain anxious for the fate of their country confronted by the inefficience of the old fighting methods which the progress of industry out of date renders daily.

By sounding the alarm in both our countries, I find myself in company with my faithful friend Field Marshal Sir Henry Wilson. This is an old habit, contracted by both of us, many years ago, which we still maintain at the present time to insure for ourselves once again, peace in the future.

Together, we say, read this work of Major Lefebure.

F. FOCH.

CONTENTS

9

Contents

Contents

Contents

ILLUSTRATIONS

13

INTRODUCTION

No one who has the welfare of the country at heart can fail to share Major Lefebure's anxiety that a clear, accurate, and unbiased account of chemical warfare should be presented to the public, so that the many erroneous ideas now prevalent in regard to poison gas and its uses may be dispelled.

The whole subject of chemical warfare is at present *sub judice,* and there is great danger that the future safety of this country may be jeopardised by the almost universal ignorance of the peculiarities and potentialities of this class of warfare. Recent publications in the Press have shown a tendency to deal with the subject on purely sentimental grounds, and attempts have been made to declare this form of warfare illegitimate without full and careful consideration of all the facts and their significance for the future.

Major Lefebure has therefore attempted in his book to make it quite clear that no convention, guarantee, or disarmament safeguard will prevent an unscrupulous enemy from employing poison gas, especially if that enemy has discovered some new powerful agent, or possesses, as Germany does in her well-organised and strong chemical industry, a ready means for producing such chemicals in bulk at practically a moment's notice; further, that the safety of this country makes it imperative that the study and investigation of the subject should be continued and that our chemical and

15

Introduction

dye industry should be developed, so that when an emergency arises we may have the necessary facilities for supply ready to hand.

It is not for me to express any opinion here either as to the desirability of using gas as a weapon or as to the possibility of preventing an enemy from using it. But I am convinced that a decision come to without full knowledge of the facts may involve grave danger and heavy preventable loss of life. I am further convinced that Major Lefebure, by his special knowledge and long experience as chemical liaison officer during the war, is well qualified to speak, and that his opinion is entitled to full consideration. For these reasons I think that his book will do a much needed public service. I wish it every success, and the greatest possible number of readers.

HENRY WILSON, F.M.

CHAPTER I

EXPLANATORY

The Riddle of the Rhine.—The Great War challenged our very existence. But with the tension released, and the Allies victorious, the check to the German menace appears crushing and complete. Few realise that one formidable challenge has not been answered. Silently menacing, the chemical threat remains unrecognised. How, asks the reader, can this be? Are we not aware of the poison gas campaign? Indeed, we have not yet grasped the simple technical facts of the case, and these are merely the outward signs of a deep-rooted menace whose nature, activities, and potentialities are doubly important because so utterly unsuspected by those whom they most threaten.

How many of us, for example, realise that the Germans relied mainly on gas for success in the great March assault of 1918, which threatened to influence the destinies of the world. Yet Ludendorff goes out of his way to tell us how much he counted upon it. How many understand that the 1918 hostilities were no longer a war of explosives. German guns were firing more than fifty per cent. of gas and war chemical. But a deep study of such war facts reveals a much more significant matter.

All are aware of the enormous national enterprises built to fulfil our explosives programme.

The Riddle of the Rhine

With mushroom-like growth chemical establishments of a magnitude hitherto unknown in England arose to meet our crying needs. What was the German equivalent, and where were the huge reservoirs of gas and war chemical which filled those countless shells? Krupp, of Essen, loomed large in the mind of every Allied citizen and soldier. There lay the sinews of war in the making. But the guns were useless without their message. Who provided it? A satisfactory answer to this question demands an examination of the great German I.G., the Interessen Gemeinschaft, the world power in organic chemical enterprise, whose monopoly existence threatened to turn the tide of war against us. This organisation emerges from the war with renewed and greater strength. Our splendid but improvised factories drained the vital forces of the nation, and now lie idle, while German war chemical production fed new life blood and grafted new tissue to the great pre-war factories of the I.G., which, if she will, she can use against us in the future. I do not claim that this German combine has at present any direct economic or military policy against world peace. In any case, the facts must speak for themselves. But the following pages will prove that the mere existence of the complete German monopoly, represented by the forces of the I.G., however free from suspicion might be the mentality and morals of those directing its activities, constitutes, in itself, a serious menace. It is, if you will, a monster camouflaged floating mine in the troubled sea of world peace, which the forces of reconstruction have left unswept. The existence of this giant monopoly raises vital military and economic questions, which are, indeed, "The Riddle of the Rhine."

Impersonal Examination of Fact.—In a sound

examination of the subject it becomes necessary to examine the activities of our former enemies very closely. Even adopting a mild view of the case, their reputation has not been unattacked, and is not left untarnished. We, however, have no desire to renew such attacks, but we wish our statement to be coldly reliable. National and international issues are at stake which require a background unprejudiced by war emotion.

Placed in a similar predicament, in reporting to his Government of the methods of German economic aggression in the United States of America, Mr. Mitchell Palmer, the Alien Property Custodian, expressed himself as follows:

"I do not advocate any trade boycott out of spirit of revenge or in retaliation for injuries done to the United States. I do not want to continue the war after the war. I am for peace. I believe that the great overshadowing result which has come from this war is the assurance of peace almost everlasting amongst the peoples of the earth. I would help to make that an absolute certainty by refusing to permit Germany to prosecute a war after the war. The military arm of her war machine has been palsied by the tremendous hammering of the allied powers. But her territory was not invaded, and if she can get out of the war with her home territory intact, rebuild a stable government, and still have her foreign markets subject to her exploitation, by means no less foul and unfair than those which she has employed on the field of battle, we shall not be safe from future onslaughts different in methods, but with the same purpose that moved her on that fateful day in July when she set out to conquer the world."

Ours is a fair standpoint. Let us know the facts of the chemical war into which Germany impelled us.

19

The Riddle of the Rhine

Let us examine its mainsprings, in conception and action, see how far they can be explained in terms of pre-war Germany, and how far they remain ready to function in the much desired peace which they threaten. If the result be unpleasant, let us not hide our heads in the sand, but exercise a wise vigilance, choose what precautions are available and consistent with our plans for world peace.

A Critical Point in Disarmament.—Probably never before in the history of man has Disarmament figured as such a vitally urgent national and international measure. Discussions and official utterances reveal a very disquieting tendency.

When compared with the methods, armament and materials of the war in 1914, those of 1918 reveal basic changes which a hundred years of former peace could not have brought about. These developments are not merely of fact, but they represent the opening of new fields, visions of possibilities previously undreamed of by the practical soldier. By the concentrated application of electricity, chemistry, and other sciences to war two dominating factors have emerged, whose importance to war, and danger for world peace, can only gain momentum with time. The scientific or technical initiative, the invention of a deadly new chemical, wireless-directed aeroplane, or other war appliance and their incidence on war through large scale production in the convertible industries of peace constitute a challenge which, if unanswered by practical schemes for world disarmament, will render the latter worse than useless, by aggravating the danger of sudden decisive attack in an otherwise disarmed world.

There is a tendency to ignore this aspect of disarmament. We appear to be thinking in terms of a world still organised for war on 1914 lines. The

20

disbanding of the German army and semi-military organisations, and the reduction of her artillery and small arms seem to occupy all our attention. Such, it might be urged, is the immediate need; we can leave the future to find answers to the other problems. This answer is dangerous, for it ignores the disarmament aspect of what is perhaps the most important development in the modern offensive campaign. We refer to poison gas or chemical warfare. This, the crux of all disarmament, is dealt with at some length in the chapters which follow.

A curiously illogical attitude of mind has arisen in certain quarters. There is a tendency among strong adherents to the ideal of world peace to regard themselves as its sole possessors. Every thinking civilian and soldier must adhere to such an ideal; the only point at issue is the method of approaching it. The mere fact that a League of Nations is called into being to attain world peace implies recognition of the fact that a definite mechanism and definite measures are required for the purpose; this is self-evident. There are those who, having established their League of Nations, feel that they can attain chemical peace by merely prohibiting chemical war, in other words, they expect their mechanism to achieve its object without functioning, to attain peace by its mere existence. Just as special measures are required to control disarmament in the older branches of warfare, in the same way special measures, but not the same measures, are required to control the chemical peace. Chemical peace guaranteed by a mere signature is no peace at all.

In a recent Press utterance we find an appeal to prohibit chemical warfare and to "trust the general sentiment of the civilised world to say that the lesson has been learnt in that sense." "There is the League

of Nations to furnish that sentiment with a mouthpiece and a sanction." We agree, but to stop there
is dangerous, the most important thing which it must
furnish is a mechanism of control, a check, or guarantee. This question is one of the most important
which confronts us for world peace. It merits the
most careful consideration.

Even responsible and relevant officials who admit
that their League must do more than issue edicts,
that their mechanism must function, are ignoring
the specific technical aspect of the war methods
whose use we wish to limit. This matter will receive
later attention.

The following pages, therefore, are an attempt to
represent the salient points in the development of
chemical warfare, its causes, results, and future.
Such an attempt cannot limit itself to merely British
developments, and this is not a final detailed memoir
of British chemical warfare. Further, in considering the future, we examine another aspect of chemical warfare. Facts lead us to believe that it was
purely the most open and obvious activity in a whole
campaign of chemical aggression which had effective
unity of conception and direction long before the
war started.

Need for a Balanced View of Chemical Warfare.
—The facts of chemical warfare have probably been
less ventilated than those of any other important
war development. Yet no subject has aroused more
general and intense feeling. Tanks, aircraft, the
different campaigns, enemy memoirs, and a variety
of war subjects, have received a considerable measure of publicity, some more than full measure.
Grave questions are pending in which the chemical
aspect of national defence is a prominent factor.
However willing the individual concerned, he cannot

make a sound judgment on the brief technical or popular garbled versions which have appeared. One searches in vain for balanced and detailed statements on the question. This may be due in no way to lack of intention, but to lack of opportunity. Therefore, no excuse is needed for this contribution, but rather an apology for the obscurity which has so far surrounded the subject. What is the cause of this emotional or almost hysterical background from which a clear definition of the matter is only now beginning to emerge? Circumstances are to blame; the first open act of chemical warfare decided the matter.

This event, the first German cloud gas attack at Ypres, arriving at the peak of allied indignation against a series of German abuses, in particular with regard to the treatment of prisoners, left the world aghast at the new atrocity. Further, its use against entirely unprotected troops was particularly revolting. The fact that such a cloud of chlorine would have passed the 1918 armies untouched behind their modern respirators, could not be known to, nor appreciated by the relatives of the 1915 casualties. But the emotion and indignation called forth by the first use of gas has survived a period of years, at the end of which the technical facts would no longer, of themselves, justify such feeling. We would hesitate to do anything which might dispel this emotional momentum were we not convinced that, unaccompanied by knowledge, it becomes a very grave danger. If we felt that the announcement of an edict was sufficient to suppress chemical warfare we would gladly stimulate any public emotion to create such an edict. But therein lies the danger. Owing to certain technical peculiarities, which can be clearly revealed by examination of the facts, it is impossible to suppress chemical warfare in this way. As well try to suppress disease by forbidding its recurrence. But we can take precaution against disease, and the following examination will show clearly that we can take similar precautions against the otherwise permanent menace of chemical war. Further, backed by such precautions, a powerful international edict has value.

It is, therefore, our intention to present a reasoned account of the development of poison gas, or chemical warfare, during the recent war. But to leave the matter there would be misleading and culpable, for, however interesting the simple facts of the chemical campaign, they owed their being to a combination of forces, whose nature and significance for the future are infinitely more important. The chief cause of the chemical war was an unsound and dangerous world distribution of industrial organic chemical forces. Unless some readjustment occurs, this will remain the "point faible" in world disarmament. We, therefore, propose to examine the relationships between chemical industry, war, and disarmament.

Some Preliminary Explanation.—The chemistry of war, developed under the stress of the poison gas campaign, is of absorbing chemical and technical interest, but it has none the less a general appeal. When its apparently disconnected and formidable facts are revealed as an essential part of a tense struggle in which move and counter-move followed swiftly one upon the other, its appeal becomes much wider. Therefore, in order not to confuse the main issue in the following chapters by entering upon tiresome definitions, it is proposed to conclude the present chapter by explaining, simply, a number of chemical warfare conceptions with which the expert is probably well acquainted.

24

"Poison Gas" a Misleading Term.—Poison gas is a misleading term, and our subject is much better described as "chemical warfare." Let us substantiate this by examining briefly the types of chemicals which were used. In the first place they were not all gases; the tendency during the war was towards the use of liquids and solids. Even the chemicals which appeared as gases on the field of battle were transported and projected as liquids, produced by compression. As the poison war developed, a large number of different chemicals became available for use by the opposing armies. These can be classified, either according to their tactical use, or according to their physiological effects on man.

The British, French, American, and German armies all tended to the final adoption of a tactical classification, but the French emphasised the physiological side. Let us use their classification as a basis for a review of the chief chemicals concerned.

The French Physiological Classification;—Asphyxiating Substances; — Toxic Substances. — Chemicals or poison gases were either asphyxiating, toxic, lachrymatory, vesicant, or sternutatory. It is perfectly true that the asphyxiating and toxic substances, used during the war, produced a higher percentage of deaths than the other three classes, but the latter were responsible for many more casualties. The so-called asphyxiating gases produced their effect by producing lesions and congestion in the pulmonary system, causing death by suffocation. The best known substances of this type was chlorine, employed in the liquid state in cylinders on the occasion of the first German gas attack, but the most formidable were phosgene (an important substance required in the manufacture of dyes), diphosgene, chlor-picrin, made from bleaching powder and picric acid, brom-acetone, which was also a powerful lachrymator, and diphenylchlorarsine, known as sneezing gas, the first sternutatory or sneezing compound to appear on the front in large quantities. The toxic compounds were so called because of their specific effect upon particular parts of the organism such as, for example, the nervous system. The chief example, with regard to the military value of which there has been much dispute, was prussic, or hydrocyanic, acid. The French had definite evidence of the mortal effect of this compound upon German gunners, but it was doubted by other Allies whether French gas shell produced a sufficient concentration of gas to be of military value. It was a kill or cure compound, for recovery was rapid from any concentration which did not produce death.

A prominent Cambridge physiologist, in the heat of the controversy on this matter, made a very brave and self-sacrificing experiment. He entered a chamber of prussic acid which was sufficiently concentrated to cause the death of other animals which were present. They were removed in time, and he escaped because the concentration was not a mortal one for man. This was, in a sense, an *experimentum crucis* and, although it did not disprove the extreme danger of prussic acid, if employed in high concentrations, it showed, on the other hand, that it was difficult to gauge the military value by field experiments; battle results were necessary. The Germans' disappointment with the use of arsenic compounds confirms this need for battle evidence.

Lachrymators.—There is hardly need to dwell on the next class, the lachrymator. These compounds were employed on a large scale to produce temporary blindness by lachrymation, or weeping. We give later some interesting examples of their use on the front. It is an arresting thought that even as early as 1887 Professor Baeyer, the renowned organic chemist of Munich, in his lectures to advanced students, included a reference to the military value of these compounds.

Vesicant or Blistering Compounds.—It was the introduction of the fourth, the vesicant class, which revealed, more than any other enemy move, the great possibilities inherent in chemical warfare. These compounds, the chief of which was mustard gas, produced vesicant, or skin burning, effects, which, although rarely mortal, were sufficient to put a man out of action for a number of months. Mustard gas resulted from pure scientific investigation as early as 1860. Victor Meyer, the famous German chemist, described the substance in 1884, indicating its skin-blistering effects. There is evidence of further investigation in German laboratories a year before the outbreak of war, and whatever the motive for this work, we know that mustard gas must have received the early attention of the German War Office, for it was approved and in production early in 1917. Although the Medecin aide-major Chevalier of the French services drew attention to its importance in 1916, the French had no serious thought of using mustard gas, and did not realise its possibilities until the German battle experiment of July, 1917. It is not generally known, however, that other vesicant compounds were employed, notably some of the arsenic compounds, and the Germans were researching on substances of this nature which gave great promise of success. Mustard gas provides a striking example of the organic way in which chemical warfare is bound up with the dye industry. The compounds required for its manufacture were those which had been made on a large scale by the I.G. for the production of indigo. World indigo monopoly meant possession of a potential mustard gas surprise on the outbreak of war.

Sneezing or Sternutatory Substances.—The last class, the sternutatory substances, produced the familiar sneezing effect which was accompanied by intense pain and irritation of the nose, throat, and respiratory channels. They were mostly arsenic compounds and were not only sternutatory but also toxic, producing the after effects of arsenic poisoning.

The Tactical Classification.—From the point of view of our account of chemical warfare, however, the physiological classification of these substances is not so important as the tactical and, indeed, once this grouping of the substances is understood, a profound knowledge of their chemical nature is not necessary.

Persistent Substances.—Two main classes exist from the tactical point of view. There are those "persistent" substances which remain for a long time on the soil or on the object on which they are sprayed by shell, while retaining their dangerous effect. Mustard gas was the chief example, but some of the lachrymators were just as persistent. By their use it is possible to render ground uninhabitable or ineffective for military movement. The combination of the vesicant and persistent properties of mustard gas rendered it a powerful military factor.

Non-Persistent Substances.—On the other hand, there are the relatively volatile substances, such as phosgene, which can be used immediately before an attack. The chief sternutatory compound, diphenylchlorarsine, although not volatile, could also be used in this way, for, being a solid and in a very finely pulverised state, its presence on the ground was not a distinct danger, and it invited chemical decomposition.

28

Penetrants.—The Germans introduced an additional tactical group. This comprised pulverised substances able to penetrate the mask on account of their existence as minute particles. The Germans expressed these tactical conceptions by their shell markings. The familiar Green Cross represented the slightly persistent, volatile, lethal compounds, such as phosgene and diphosgene. The German gunner had no need to know the content of his gas shell so long as he could identify the cross. Yellow Cross, representing mustard gas, was the most highly persistent type. It is interesting to speculate whether a new persistent compound, whose military value was due to some other property than the blistering, would have been grouped under Yellow Cross. Logically, this should have been done. Blue Cross covered the arsenic group of compounds, which were non-persistent and were expected to penetrate the mask. So strong was this tactical conception that the Allies were on the verge of adopting a uniform shell marking based on this principle throughout their armies.

Special Gas Weapons and Appliances.—It is a popular misconception that gas was only discharged from cylinders in huge clouds, or used as artillery shell. A number of special weapons developed, which were particularly adapted for gas. Thus, the Livens projector, which was a great Allied advance, produced a gas cloud a long distance from the point of discharge, while the Stokes and other short range guns were used for rapid fire of large numbers of gas shell.

The primary conceptions with regard to protection have been brought home to so many, through the fact that the mask was a part of the equipment of every soldier, that we need not dwell on them here. It is not generally realised, however, that every modification introduced by either side was a vital and direct counter to some enemy move planned to render the protection of the opponent ineffective.

Gas Shell.—A word is necessary to define the use of gas shell. The point which must be realised is that gas, and in particular gas shell, fulfilled a special purpose in warfare, from which it was much more suitable than explosives. The use for neutralising batteries, cross roads, and rendering whole areas uninhabitable, is developed fully in our reference to the great German attacks in 1918.

With this brief sketch to clear the ground, we can embark more freely upon the account of chemical warfare which follows.

To face page 30

TYPICAL GAS SHELL BURSTING.

CHAPTER II

Ypres, April, 1915, to the Somme, August, 1916.

The First Cloud Gas Attack.—The critical factor of surprise in war was never nearer decisive success than on April 22nd, 1915. Of this, the occasion of the first German gas attack at Ypres, Field-Marshal Sir J. D. P. French stated:

"Following a heavy bombardment, the enemy attacked the French Division at about 5 p.m., using asphyxiating gases for the first time. Aircraft reported that at about 5 p.m. thick yellow smoke had been seen issuing from the German trenches between Langemarck and Bixschoote. What follows almost defies description. The effect of these poisonous gases was so virulent as to render the whole of the line held by the French Division mentioned above practically incapable of any action at all. It was at first impossible for any one to realise what had actually happened. The smoke and fumes hid everything from sight, and hundreds of men were thrown into a comatose or dying condition, and within an hour the whole position had to be abandoned, together with about fifty guns. I wish particularly to repudiate any idea of attaching the least blame to the French Division for this unfortunate incident."

31

The Element of Surprise.—The enemy just missed colossal success rendered possible by the use of an entirely new war method; one contrary to engagements entered into by them at the Hague Convention.

There were elements in this first gas attack which were absent even from the situation created by our first use of tanks. Unfamiliarity amongst the troops, or the staff, for that matter, created an atmosphere of unparalleled confusion. Men attempted to protect themselves by burying their mouths and nostrils in the loose earth. Those chemists, on the spot, not immediately struck down, made frantic efforts to bring up supplies of any suitable and available chemical or material which might assist resistance and movement in the affected zone. Paying every homage to the heroic sacrifices and brave actions which characterised the Allied resistance, we cannot ignore the fact that morale must have been very severely shaken locally, and that a general disquiet and uneasiness must have permeated the whole front until measures were known to be effectively in progress, not only for protection, but for retaliation. The enemy had but to exploit the attack fully to break through to the channel ports, but failed to do so. The master mind behind this new and deadly attack was not, let us remember, that of a soldier. It was very strongly rumoured that this monstrous conception and its execution were due to one or, at the most, two renowned German Professors. The first hammer blow in the enemy chemical campaign was a two-party conspiracy, led by world-famous scientists and the powerful I.G. with the German army unconvinced but expectant, little more than a willing dupe.

The German Surprise

Lord Kitchener's Protest.—In his spirited protest in the House of Lords, Lord Kitchener stated: "The Germans have, in the last week, introduced a method of placing their opponents *hors de combat* by the use of asphyxiating and deleterious gases, and they employ these poisonous methods to prevail when their attack, according to the rules of war, might have otherwise failed. On this subject I would remind your Lordships that Germany was a signatory to the following article in the Hague Convention:

" 'The Contracting Powers agree to abstain from the use of projectiles the object of which is the diffusion of asphyxiating or deleterious gases.' "

This protest circulated amongst neutrals prompted numerous attempts at vindication in the German Press. In several cases we find important newspapers arguing that the German attack was not contrary to the Hague Convention, while others admitted the breach, but claimed that the Germans merely followed Allied example. The main technical excuse was that the effect of the German gas was merely stupefying (*Colniche Zeitung*, June, 1915). It is incredible that the German nation was, or could allow itself to be, so hoodwinked. Scientific Germany was certainly aware of the true nature of the gases used. Even scientific neutrals in Berlin at the outbreak of war, and during the ensuing winter, were aware of the German poison gas work, which commenced, in an organised way, almost as soon as war broke out. The Germans have argued that they only entertained the idea of gas after Allied use. The facts revealed below are a sufficient answer. Whatever legal arguments may be

33

involved, there is no doubt as to German intention.

We do not wish to enter into a comprehensive examination of the legal aspect of the first use of cloud and shell gas by Germany. Whatever complicated arguments may turn upon the strict reading of a phrase in the records of the Hague Convention, we have no doubt whatever as to the desires and intentions of the Assembly, and we regard Germany (and the Allies) as morally engaged not to venture upon the series of chemical enterprises which she openly commenced with the Ypres cloud attack. The Versailles Treaty also renders fruitless any such discussion. Article 171, accepted by Germany, is deliberately based on her breach of International Convention.

German Preparations.—A significant phrase occurs in the Field-Marshal's despatch. "The brain power and thought which has evidently been at work before this unworthy method of making war reached the pitch of efficiency which has been demonstrated in its practice shows that the Germans must have harboured these designs for a long time." This is a most important point. It was argued by many generous and fairminded people in April, 1915, that the German use of gas was the result of a sudden decision, only arrived at in a desperate effort to terminate the war. This point of view would give us maximum hope for the future. But the actual truth? What do we know about German preparations, and how far back do they date? Any preparations which occurred must have covered research on the compounds to be employed and on the protection required for the German troops, their training for the cloud attack and the design and production of the special appliances to be used.

Finally, the production of the chemicals themselves had to be faced.

Research.—We have obtained an insight into the German research preparations, which leaves no doubt as to their intention. There is evidence that the Kaiser Wilhelm Institute and the physico-chemical institute near by were employed for this purpose as early as August, 1914. Reliable authority exists for the statement that soon after this date they were working with cacodyl oxide and phosgene, both well known before the war for their very poisonous nature, for use, it was believed, in hand grenades. Our quotations are from a statement by a neutral then working at the Institute. "We could hear the tests that Professor Haber was carrying out at the back of the Institute, with the military authorities, who in their steel-grey cars came to Haber's Institute every morning." "The work was pushed day and night, and many times I saw activity in the building at eleven o'clock in the evening. It was common knowledge that Haber was pushing these men as hard as he could." Sachur was Professor Haber's assistant. "One morning there was a violent explosion in the room in which most of this war work was carried out. The room was instantly filled with dense clouds of arsenic oxide." "The janitors began to clear the room by a hose and discovered Professor Sachur." He was very badly hurt and died soon after. "After that accident I believe the work on cacodyl oxide and phosgene was suspended and I believe that work was carried out on chlorine or chlorine compounds." "There were seven or eight men working in the Institute on these problems, but we heard nothing more until Haber went to the Battle of Ypres." Rumours to this effect circulated in 1915.

The Riddle of the Rhine

Production.—Preparations for production can easily be imagined. The Germans first used chlorine for cloud gas, and certain lachrymators for shell. The chlorine was readily available. At about this time British liquid chlorine capacity had a maximum daily output of about one ton, while along the Rhine alone the production was more than forty times greater. The question of German chlorine production was, therefore, already solved. The lachrymators were mainly raw materials and intermediates of the dye industry submitted to a process, the technique of which the German dye factories readily mastered. Here, again, production presented no real difficulties. Cylinders were also probably available from the industry.

Field Preparations.—There remains the last question of gas attack technique and personnel. Those of us who remember the difficulties involved in creating our own organisation in the summer of 1915 have no illusions on the question of German preparation. Giving the Germans every credit for their technical and military efficiency, some months must have been occupied in establishing and training the special companies required, and in arriving at a satisfactory design for the discharge appliances. Schwarte's book, *Die Technik Im Weltkriege,*[1] tells us "specially organised and trained troops" were required for the purpose. Prisoners taken later revealed the German methods. Gas officers and N.C.O.'s, after making a careful survey of the front line trench, organised the digging of deep narrow trenches at suitable places below the surface of the main trench, just underneath the parapet. The heavy gas cylinders, weighing as much as ninety pounds, were carried to the front line by the un-

[1] *Die Technik Im Weltkriege.* Publisher: Mittler, Berlin, 1920.

fortunate infantry. The discharge valves were carefully protected by domes which screwed on to the cylinder. The latter were introduced into the holes, tops flush with the trench bottom, and covered by a board on which reposed the "Salzdecke," a kind of long bag stuffed with some such material as peat moss and soaked in potash solution to absorb any slight gas leakages. Three layers of sandbags were built above the salzdecke to protect the cylinder from shell fragments and to form a firestep for the infantry. This concealed the cylinders so efficiently that, in our own trenches, I have often found the new occupants of a sector ignorant of the presence of gas cylinders under their own firesteps. On the favourable night the dome was removed and a lead pipe was connected to the cylinder and directed over the parapet into No Man's Land, with the nozzle weighed down by a sandbag. The pioneers stood by the batteries of twenty cylinders each and let off the gas a fixed few minutes after a rocket signal, at which the infantry retired to leave the front line free for the pioneers, who not only ran the risk of gassing from defective appliances but were subjected to almost immediate violent bombardment from the opposing artillery. When surprise was complete artillery retaliation was very late in developing. This gives a faint idea of the elaborate preparations required. They must have been doubly arduous and lengthy on the very first occasion of cloud gas attack.

German Opinion of Results.—We can now regard the chlorine attack of April 22, 1915, as the first and successful result of a huge German experiment on a new method of war, the pioneer work of which actually began at (if not before) the outbreak of war. Quoting again from Schwarte:

"G.H.Q. considered the attack near Ypres to be a successful experiment. The impression created was colossal and the result not inconsiderable, although it was not fully utilised from the tactical point of view. It was obvious that we had gained a great advantage; the enemy was not sufficiently prepared with defensive measures against gas." Indeed, we were absolutely unprepared, so much so, that after the German attack nearly every household in England contributed to our first inefficient and improvised mask. Is not this suggestion of our preparation a deliberate attempt to deceive the German public? They seem to have been as easily hoodwinked on gas questions as on many others.

Germany Prompted by Production Monopoly. —An important point arises. The Germans failed to exploit their initial success. This is not very surprising. Whatever the opinion of the chemists behind the movement, the German General Staff must have retained the elements of precaution in its opinion. It could not have taken for granted the formidable success which the chemists proved justified in prophesying. This being so, we can fairly assume that had there been very serious difficulties in carrying out this huge war experiment it might never have materialised. Such difficulties might have been found in production. But as we have seen, the question of production was the most easily forged link in the chain of events which led to the use of poison gas by Germany. In other words, this monopoly in ease of production was an inducement to the Germans to proceed with their experiment.

The earliest German cloud gas attacks established beyond a doubt the enormous value of gas against unprotected troops, in other words, its value as a complete surprise. These conditions were again approached in the first German use of mustard gas. The most telling examples will probably be found in the future, unless the correct precautions are taken. The whole history of chemical warfare during the war was a struggle for this initiative, a struggle between gas protection and aggression.

Standard Uses for Gas;—Gas Shell.—But gas found an important use besides that of strategic surprise. It became a standard weapon for certain clear and definite tactical purposes. (For some of these, indeed, the factor of local surprise was important.) We refer to the specific use of gas shell for the neutralisation of batteries, roads, and areas, and to the use of cloud gas, prior to offensives for the production of casualties, and wearing down of reserves. The Ypres attack had not by any means established the use of gas for such purposes. There is no doubt that, from this point of view, the experimental period carried on for many months. Naturally, in some respects, there was always an experimental element in the use of gas.

Further German Cloud Attacks.—Two days after the first cloud gas attack the Germans launched a second against the Canadians, with similar results. Quoting from official despatches: "On the early morning of the 24th a violent outburst of gas against nearly the whole front was varied by heavy shell fire, and a most determined attack was delivered against our position east of Ypres. The real attack commenced at 2.45 a.m. A large proportion of the men were asleep, and the attack was too sudden to give them time to put on their respirators." These latter were hurriedly improvised after the first Ypres attack.

Hill 60.—Four more attacks occurred in May, notably in the region of Hill 60. "On May 1st another attempt to recapture Hill 60 was supported by great volumes of asphyxiating gas which caused nearly all the men along a front of about 400 yards to be immediately struck down by its fumes." "A second and more severe gas attack under much more favourable weather conditions enabled the enemy to recapture this position on May 5th. The enemy owes his success in this last attack entirely to the use of asphyxiating gas." "It was only a few days later that the means which have since proved so effective of counteracting these methods of making war were put into practice." (Official despatches, 1915.) The despatch further described how violent bombardments, the confusion and demoralisation from the first great gas surprise, and subsequent almost daily gas attacks, prevented the proper reorganisation of the line in question.

Origin of German Gas Shell.—After May a long period elapsed during which the Germans confined their war chemical activities on the front to the use of gas shell. Schwarte's book describes their origin as follows:—"The main idea which influenced the *first* construction of a German projectile containing chemicals (October, 1914) was that of adding to the charge an irritant substance, which would be pulverised by the explosion of the projectile, and would overwhelm the enemy with a cloud of dust. This cloud would hover in the air and have such an effect upon the mucous membranes that, for the time being, the enemy would be unable to fight in such an atmosphere. By altering the construction of the 10.5 c.m. universal shell for light field howitzers, the 'N.i' projectile was created in the form of 10.5 c.m. shrapnel, the bullets of which were embedded in a sternutatory powder (double salts of dianisidine) well stamped down, instead of an explosive. By means of the propelling charge and the grinding effect of the bullets, this powder was pulverised on explosion. The irritation caused was not very intense, lasted only a short time and affected only a limited area and therefore it was of no importance in the field, but the initial step had been taken. Liquid irritants soon came to the front—xylyl bromide and xylylene dibromide—a mixture used later under the name of T. stuff, bromo-acetone and brominated methyl ethyl ketone, later introduced under the name of B. stuff and Bn. stuff."

During experiments they gave such improved results in intensity, in power of lasting and of affecting an increased area, that practical results in the field were ensured. The use of these liquids in projectiles, however, was contrary to the accepted idea with regard to artillery, according to which liquid materials should not be used for ballistic reasons. Specially arranged shoots were required to prove that the projectiles in use in the German Army could also be used from the ballistic point of view when filled with liquids.

In this way the first effective German gas projectile, the T. shell for heavy field howitzers, was evolved (January, 1915).

Early German Gas Shell.—The first important use of German gas in shell was that of brominated and chlorinated organic compounds, T. and K. stuffs. Schwarte's book tells us "the use of these projectiles was continually hampered by lack of understanding on the part of the troops which it was difficult to overcome. In the summer of 1915 it was practically in the Argonne alone that any considerable results were attained by the new projectiles." And he describes how the first elements of the new gas tactics were developed there.

A Successful Experiment.—The development of the gas shell, the use of which, generally speaking, is independent of, but co-ordinated with, wind direction, may have received stimulus from the fact that the prevailing wind, so important for cloud gas, favoured the Allies. It is clear that this period was an experimental one, but we know that by August, 1915, Geman military opinion had crystallised out to the extent of formulating certain rules, issued as Falkenhayn's orders for the employment of gas shell. These early orders defined two types of shell, one persistent, for harassing purposes, and the other non-persistent, to be used immediately before an attack. They specified the number of shell to be used for a given task. But in this they were unsound and it is clear that the Germans had an exaggerated opinion of what could be achieved with a small number of shell. They adhered too closely to high explosive practice. Various documents reveal the fact that the Germans were much more satisfied with their gas tactics than they would have been had they possessed information with regard to our losses from their shell. They attached insufficient importance to the value of surprise and highly concentrated shoots, and had a mistaken idea of the actual specific aggressive value of their early types.

Lachrymators at Loos, 1915.—Germany commenced the manufacture of lachrymators, crude brominated xylene or brominated ketones, early in, or perhaps before 1915. These substances caused great inconvenience through temporary blindness by lachrymation, but were not highly toxic. In June, 1915, however, they began to produce lethal gas for shell. Falkenhayn's orders for the use of gas shell, mentioned above, although they represent by no means the best final practice, were definite evidence that gas had come to stay with the Germans. The writer has vivid recollections of their use of lachrymators in the Loos Battle. Batteries in the open, under the crest near the Lens road, were in position so that the wind direction practically enfiladed them, sweeping along from the direction of Le Rutoire farm. Gas from German shell, borne on the wind, was continually enveloping the line of batteries, but they remained in action. It was on this occasion while watching the bursting gas shells from the outskirts of the mining village of Philosophe that Major-General Wing was killed outright by a high explosive shell. These gas shells certainly did not achieve the results which the Germans expected, although they were not without effect. Demolished villages, the only shelter for troops in a desolate area, have been rendered uninhabitable for days by a concentrated lachrymator enemy shoot of less than one hour. Again, walking into gas "pockets" up a trench one has been stopped as by a fierce blow across the eyes, the lachrymatory effect was so piercing and sudden. The great inconvenience which was occasioned to parties engaged in the routine of trench warfare, on ration or engineering duties, and the effect on movement in the rear after an assault, taken cumulatively, represented a big military factor.

The Flammenwerfer.—There can be no doubt that this period marks increasing German willingness to live up to their "blood and iron" theories of war, and, in July, 1915, another device with a considerable surprise value was used against us: the flame projector, or the German flammenwerfer. Field-Marshal Sir John French signalled the entry of this new weapon as follows: "Since my last despatch a new device has been adopted by the enemy for driving burning liquid into our trenches with a strong jet. Thus supported, an attack was made on the trenches of the Second Army at Hooge, on the Menin Road, early on 30th July. Most of the infantry occupying these trenches were driven back, but their retirement was due far more to the surprise and temporary confusion caused by the burning liquid than to the actual damage inflicted. Gallant endeavours were made by repeated counter-attacks to recapture the lost section of trenches. These, however, proving unsuccessful and costly, a new line of trenches was consolidated a short distance farther back."

Although this weapon continued to be used right through the campaign, it did not exert that influence which first acquaintance with it might have led one to conclude. At the same time, there exists a mistaken notion that the flame projector was a negligible quantity. This may be fairly true of the huge non-portable types, but it is certainly not true of the very efficient portable flame projector which was the form officially adopted by the German, and later by the French, armies. On a number of occasions Germany gained local successes purely owing to the momentary surprise effect of the flame projector, and the French made some use of it for clearing out captured trench systems over which successful waves of assault had passed. Further, the idea of flame projection is not without certain possibilities for war.

German Phosgene Clouds.—Germany had by no means abandoned cloud gas, however. She had merely been planning to regain what the Ypres attacks had lost for her, the cloud gas initiative. We have seen how phosgene had occupied the attention of the German research organisation in the first months of the war. Once alive to its great importance, they must have strained all efforts to obtain an efficient method of using it at the front. Phosgene was remarkable for its peculiar "delayed" effect. Relatively small quantities, inhaled and followed by vigorous or even normal exercise, led to sudden collapse and fatal effects sometimes more than twenty-four hours after the attack. The case of a German prisoner in a First Army raid after a British gas attack was often quoted on the front. He passed through the various Intelligence headquarters as far as the Army, explaining the feeble effect of the British gas and his own complete recovery. But he died from delayed action within twenty-four hours of his last interrogation. This effect imposed strict conditions of discipline, and men merely suspected of exposure to phosgene were compelled to report as serious casualties and carried as such even from the front line.

The successful development of the phosgene cloud probably arrived too late for the Ypres attacks, and a variety of reasons must have led to the postponement of its use until such time as it might once again give Germany the real initiative. Accordingly, on December 19, 1915, a formidable cloud gas attack was made on the north-east of the Ypres salient, using a mixture of phosgene and chlorine in a very high concentration. Fortunately, by this time we had established an anti-gas organisation, which had forestalled the production of cloud phosgene by special modifications in the British respirator. The conditions were similar to those of April 22nd, 1915. Instead of the first use of cloud gas, we had the first use of the new gas in highly concentrated cloud. In both cases the Germans reckoned on our lack of protection, correctly in the first case, but incorrectly in the second. In both cases they were sure that great difficulties in production would meet our attempts at retaliation. In general this proved true, but in this case and increasingly throughout the war, they reckoned without Allied adaptability. The French development of phosgene manufacture was indeed remarkable.

Very interesting light is thrown on this attack by Major Barley, D.S.O., Chemical Adviser to the British Second Army. It appears that in November, 1915, the French captured a prisoner who had attended a gas school in one of the factories of the I.G. Here lecturers explained that a new gas was to be used against the British forces, many thousands of casualties were expected, and an attack would follow, which, correcting the errors of the effort at Ypres, would lead to the capture of the Channel ports. Efforts were at once made to obtain information on gas preparation by the Germans in front of the British sectors. In this way a sergeant-major was captured on the morning of December 16th, and he revealed the date and front on which the cylinders were installed. About 35,000 British troops were found to be in the direct line of the gas, but owing to the timely warning and to the protection which had recently been adopted, we experienced very few casualties. The Germans had prepared a huge infantry attack, and used a new type of gas shell on this occasion. German troops massing must have received huge casualties owing to our preparation and the failure of their gas attack.

The last German cloud attack on the British front occurred on August 8, 1916. There were later attacks against the French, but the Germans were replacing the cloud method by other methods which they considered more suitable. These will be discussed later on, when considering our own reaction against the chemical offensive.

Gas and the Eastern Theatre.—The German surprise was not limited to activities on the Western front. In fact, apart from the first Ypres attack, cloud gas probably reaped more casualties in the East against Russia. We learn from Schwarte's book: "From reliable descriptions we know that our gas troops caused an unusual amount of damage to the enemy—especially in the East—with very little expenditure of effort. The special battalion formed by Austria-Hungary was, unfortunately, of no special importance for various reasons."

Had the nature of the Russian campaign been different, with a smaller front, and nearer critical objectives to the front of attack, we have no doubt that gas would have assumed enormous importance in the East. Russia, even more feebly organised for production than ourselves, would have been at a tremendous disadvantage, both from the point of view of protection and of the retention of satisfactory morale by retaliation.

Conclusion.—This, then, was the period of the German surprise, during which the first big shock occurred, and which promised most success for further attempts owing to the lack of comprehensive protection by the Allies. Looking at the matter in a very broad way, ignoring the moral and legal aspects of the case, we can describe this period as an example of brilliant chemical opportunism. According to plan or otherwise, conditions for this experiment were ripe in Germany as in no other country. Overcoming whatever prejudices may have existed, the German authorities realised this, seized the opportunity, and very nearly succeeded.

CHAPTER III

Loos, September, 1915, to Ypres, July, 1917.

The Need of Retaliation.—The conclusive sign of the Allied reaction to the German poison gas attack appeared at the battle of Loos. "Owing to the repeated use by the enemy of asphyxiating gas in their attacks on our positions," says Field-Marshal French in his despatch of October 15, 1915, "I have been compelled to resort to similar methods, and a detachment was organised for this purpose, which took part in the operations commencing on the 25th September for the first time." Five months thus elapsed before retaliation. From a military point of view their can be no doubt as to the wisdom, in fact the absolute necessity, of using gas in order to reply to the many German attacks of this nature. The question of morale was bound up in this retaliation. Had the Germans continued their chemical attacks in variety and extent as they did, and had it been realised that for some reason or other we were not able to retaliate in kind, none but the gravest consequences could have resulted with regard to morale. It must be remembered that the earlier use of cloud and shell gas by the Germans was of local incidence, when compared with its tremendous use along the whole of the front in the later stages of the war.

48

First Signs.—Our preparatory period was one of feverish, if somewhat unco-ordinated, activity. The production of a protective appliance, the gas mask, was vital. This development will be considered later. Allied chemical warfare organisations arose, to become an important factor in the later stages of the war. The history of Allied gas organisation is one of the gradual recognition that chemical warfare represented a new weapon with new possibilities, new specific uses, and new requirements from the rear. Its beginnings are seen in the English and French Scientific Advisory Committees appointed to examine the new German method. One could always trace an element of reluctance, however, in Allied development, signs that each move was forced upon us by some new German surprise. We find the other extreme, the logical outcome of war experience, in the completely independent Chemical Warfare Service now actually adopted in the United States of America. This is dealt with in a separate chapter.

The decision to retaliate once made, our difficulties commenced. We required gas, weapons, and methods for its use, trained personnel, and the association of certain scientific with military standards without losing the field efficiency of the latter. The German staff found this in their co-operation with eminent scientists, notably Professor Haber. Without drawing invidious distinctions between pre-war military and public appreciation of chemical science in England and Germany, it would be merely untrue to state that the Germans were not in a position of advantage in this respect. However, chemical mobilisation and co-operation proceeded sufficiently rapidly to provide us with personnel and material for the Loos attack.

The assembly and organisation of personnel occurred in three directions. In the first place the Royal Society had already begun to mobilise prominent scientists for other war purposes. In the second place, different formations in the field, realising the need for specialist treatment of the gas question, after the first German attack, created staff appointments for certain chemists chosen from infantry regiments and other formations on the front. Thirdly, men were collected at a depot in France to form the nucleus of the offensive gas troops. For this purpose chemists were specially enrolled and chosen men from infantry and other front line units were added. Early gas attacks and gas organisation did not appear to justify the immobilisation of so much chemical talent in the offensive gas troops, when chemists were needed all over England for munition production so vital to war. But later events justified the mobilisation and military training of these specialists. The expansion of the advisory and offensive organisations at the front necessitated a large number of officers, whose chemical training was of great value. It is difficult to see where they would have been found had they not been mobilised with the Special Companies. Moreover, their offensive and battle experience gained with the latter was of great value. Six or seven weeks' training witnessed the conversion of a few hundred men of the above type into one or two so called Special Companies. The spirit and work of these men in the Loos attack cannot be spoken of too highly.

The Loos Attack, September, 1915.—The Field-Marshal bears testimony to its success as follows: "Although the enemy was known to have been prepared for such reprisals, our gas attack met with marked success, and produced a demoralising effect in some of the opposing units, of which ample evidence was forthcoming in the captured trenches. The men who undertook this work carried out their unfamiliar duties during a heavy bombardment with conspicuous gallantry and coolness; and I feel confident in their ability to more than hold their own should the enemy again resort to this method of warfare."

There is evidence, however, that this early attack, inefficient as it appeared to be to participants, met with considerable success. Schwarte's book tells us: "The English succeeded in releasing gas clouds on a large scale. Their success on this occasion was due to the fact that they took us by surprise. Our troops refused to believe in the danger and were not sufficiently adept in the use of defensive measures as prescribed by G.H.Q."

On the occasion of a cloud attack a few weeks later, at the storming of the Hohenzollern redoubt, Sergeant-Major Dawson, in charge of a sector of gas emplacements in the front line trench, won the Victoria Cross. The German reply to our bombardment was very severe and under stress of it a battery of our cylinders, either through a direct hit or faulty connections, began to pour gas into our own trenches. In order to prevent panic and casualties among our own troops at this critical time, a few minutes before zero, the moment of assault, Sergeant-Major Dawson climbed on to the parapet under a hail of shell, rifle, and machine-gun fire, and, hauling up the cylinders in question, carried them to a safe distance into the poisoned atmosphere of No Man's Land and ensured their complete discharge by boring them with a rifle bullet. In addition to the Hohenzollern attack cloud gas was used in December, 1915, in the region of Givenchy.

The Riddle of the Rhine

The Somme Battle, 1916.—My impression as an eyewitness and participator, however, was that the real British gas offensive began after, and as a result of, the Loos experience. Material, organisation, and numbers of personnel, both at the front and at home, co-operation with staffs and tactical conceptions all improved vastly in time to contribute largely to the efficiency of preparations for the Somme offensive in July, 1916. During the early months of 1916, a Special Brigade was created by expanding the four Special Companies, and the 4-inch Stokes mortar was adopted, training being vigorously pursued. As many as 110 cloud gas discharges, mainly of a phosgene mixture, occurred during the Somme battle, and evidence of their success is seen in German reports. These successes were due not only to the magnitude of our operations, but to the carefully developed cloud attack tactics which aimed at obtaining maximum results from the gas employed. The factor of surprise governed all other considerations. Attacks occurred at night and depended for success upon the concentration of the maximum amount of gas in the given sector for a short, sharp discharge under the best wind conditions. There is abundant evidence of our success in these attacks. Probably the most marked feature of the captured documents or of prisoners' statements during the later stages of the Somme battle was the continual reference to the deadly effect of British cloud gas. The captured letter of a German soldier writing home stated: "Since the beginning of July an unparalleled slaughter has been going on. Not a day passes but the English let off their gas waves at one place or another. I will give you only one instance of this gas; men 7 and 8 kilometres behind the front line became unconscious from the tail of the gas cloud, and its effects are felt 12 kilometres behind the front. It is deadly stuff."

The accuracy of this reference to the long range effect of our gas clouds is borne out in a number of other statements. For example, we learnt from a prisoner examined by the French: "The men were thrown into disorder and raised their masks because they were suffocated. Many fell in running to the rear; a number did not become ill until the next day. Vegetation was burnt up to a depth of 8 kilometres." Again, prisoners taken at Maurepas stated that one of the English gas attacks was effective 10 kilometres back.

There are also marked references to the surprise nature of our gas attacks, which are an unconscious tribute to the successful tactical developments which have already been referred to, and also numerous other references to the "delayed" action of phosgene. The prisoner mentioned above, taken at Maurepas, gave testimony that some were only taken ill after several days, and one died suddenly two days after, whilst writing a letter. One prisoner, pointing to Les Ayettes on the map, stated that about the beginning of September when gas came over suddenly in the late evening, they thought it was from artillery fire because it was so sudden. No one was expecting gas and very few were carrying their masks. Another one stated: "The attack was a surprise and the cloud came over and passed fairly quickly. The whole thing did not occupy more than ten minutes." More than thirty per cent. of the battalion was put out of action.

Finally, to show what a serious imposition this constant cloud gas attack was upon the German Army, we will quote from the Special Correspondent of the *Vossiches Zeitung*. He said: "I devote a special chapter to this plague of our Somme warriors. It is not only when systematic gas attacks are made that they have to struggle with this devilish and intangible foe." He refers to the use of gas shell, and says: "This invisible and perilous spectre of the air threatens and lies in wait on all roads leading to the front."

In a despatch dated December 23rd, 1916, from Field-Marshal Sir Douglas Haig, G.C.B., the situation is ably summarised: "The employment by the enemy of gas and of liquid flame as weapons of offence compelled us not only to discover ways to protect our troops from their effects but also to devise means to make use of the same instruments of destruction. Great fertility of invention has been shown, and very great credit is due to the special personnel employed for the rapidity and success with which these new arms have been developed and perfected, and for the very great devotion to duty they have displayed in a difficult and dangerous service. The army owes its thanks to the chemists, physiologists, and physicists of the highest rank who devoted their energies to enable us to surpass the enemy in the use of a means of warfare which took the civilised world by surprise. Our own experience of the numerous experiments and trials necessary before gas and flame could be used, of the preparations which had to be made for their manufacture, and of the special training required for the personnel employed, shows that the employment of such methods by the Germans was not the result of a desperate decision, but had been prepared for deliberately.

"Since we have been compelled, in self-defence, to use similar methods, it is satisfactory to be able to record, on the evidence of prisoners, of documents captured, and of our own observation, that the enemy has suffered heavy casualties from our gas attacks, while the means of protection adopted by us have proved thoroughly effective."

One of the causes which leads to a lack of understanding of the chemical weapon is the fact that the results of chemical attack are not, like those of a huge assault, obvious to the mere visual observer. A period of months often elapsed during the war before the immediate effect of a gas attack was known. It was inspiring to witness the assault of the 18th Division near Montauban on July 1st, 1916. But few realised the part played by the preparatory gas attacks in that and other sectors of the line, in weakening the numerical strength and battle morale of effective reserves. It is, therefore, of great interest to follow up a particular case and to obtain a connected idea of the series of events associated with some particular attack.

The early stages of the Somme battle were characterised by a number of cloud gas attacks which served the double purpose of a feint, and reducing the strength of available reserves. These attacks occurred chiefly along the part of the line north of the Somme battle zone, and they extended as far as the sea. One of them occurred on the 30th August, 1916, at Monchy, between Arras and Bapaume. About one thousand cylinders were discharged during the night. The usual careful organisation preceded the attack and it is quite likely that it shared the advantage of surprise common to a large number of these attacks. Three German regiments were holding the line directly in front of the British sector concerned. Before December,

1916, the following reliable information was collected from prisoners and confirmed by cross-examination. One Company of the 23rd regiment was in training and had no gas masks with it. The gas came along quickly and about half the Company were killed. After that there were more stringent rules about carrying masks. They had no recollection of a gas alarm being sounded. Another man said that in his Company no special drill or training was being done, and a large number of men were put out of action through not being able to adjust their respirators in time. There was no warning, although after this gas alarms were given by ringing church bells. Other prisoners, from the 63rd regiment, had such vivid recollections of the attack that they said: "The effects of the English gas are said to be appalling." Collecting information from prisoners belonging to this or that Company, and carefully checking by cross-examination, it is clear that this attack must have been responsible for many hundreds of casualties.

Reasons for British Cloud Gas Success.—The fact that the British persisted with cloud gas attack and attained so much more success than the Germans, after the first surprise, was due to a curious combination of causes, quite apart from the prevailing favourable wind.

Our Casualties.—In the first place, we knew from bitter experience the deadly effect of a successfully operated cloud gas attack. We knew, for example, that in the first attack at Ypres there were more than 5000 dead with many more times that number of casualties. On the other hand, the Germans, left to speculate on our casualties, retained the conviction, from apparent non-success, that cloud gas was not a suitable form of preparation behind which to develop big infantry attacks. Quoting from Schwarte: "Large gains of ground could hardly be attained by means of an attack which followed the use of gas clouds, therefore such clouds were soon merely employed as a means of injuring the enemy, and were not followed up by an attack." This represented German policy, and it lacked vision. They did not realise that their difficulty was the method of forming the cloud, and that if a more mobile and long range method of cloud formation materialised, with correspondingly less dependence on wind direction, the object which they once sought and failed to attain would again be within their reach.

Exhausting Preparations for Cloud Attack.— The second reason accounting for the relatively early cessation of German cloud attacks is one constantly referred to in the German war memoirs. It was the enormous mechanical and muscular effort required in preparing for such an attack. Few people realise what hours of agonised effort were involved in preparing and executing a cloud gas attack. The cylinders had to be in position in specially chosen emplacements in the front line within certain time limits. The "carrying in" could not be spread over an indefinite period and usually took from two to six nights, according to the magnitude of the attack and the local difficulties. Naturally, all the work occurred in the dark. Picture the amount of organisation and labour required to install 2000 cylinders on, say, a two mile front. These cylinders would have to be assembled at a number of points in the rear of the given line where the roads met the communication trenches. No horse or lorry transport could assemble at such points before dark, nor be left standing there after dawn. To carry this number of cylinders more than fifty lorries would be required or, say, perhaps, 90 G.S. wagons. All the points of assembly would be under possible enemy shell fire. These points would be normally in use for the unloading of rations and trench engineering materials, etc., with which cylinder transport would have to be co-ordinated. Once arrived at the unloading points, parties had to be provided for unloading the lorries and for conveying the cylinders up to the front line trench. In a normally difficult trench system, for a carry of a mile to a mile and a half of communication trench, at least four men per cylinder are required to give the necessary margin for casualties and reliefs, etc. This implies the organisation of more than 8000 officers and men for the installation, with a fundamental condition that only small groups of these men be assembled at any one point at any given time. The installation of gas for an attack on this scale would have been a matter of vast and complicated organisation if there were no other activities in the trench system, and no enemy to harass the work. But to co-ordinate such an enterprise with the busy night life of the trench system and to leave the enemy unaware of your activities was a task which tried the patience, not only of the Special Companies, who organised, guided, and controlled these operations, but much more so of the Infantry Brigades and Divisions whose dispositions were interfered with, and who had to provide the men for the work.

Add to this even more acute difficulties. The front line trench is nothing but a series of traverses, thus to avoid the enfilade effect of shell and machine-gun fire. A straight trench is a death-trap. But to carry hundreds of pole-slung cylinders, already weighing as lead, round traverses on a dark night, is a feat requiring superhuman endurance. Therefore many "carries" finished with a hundred yards "over the

58

top" through the parados wire, to the near locality of the appropriate emplacement in the front line. This last carry was critical; a false step, the clatter of falling metal, meant drawing the fire of some curious and alert German machine gunner. The sudden turning of darkness into day by enemy Very lights imposed instantaneous immobility. Yet all the time tired men were straining at their heavy burden and any moment a cylinder might be pierced by intentional or unaimed rifle fire.

But the spirit of the infantry in this work, as in all they undertook, is to their everlasting credit. These tasks were an enemy challenge, and they accepted it successfully, albeit with much cursing. The work was indeed beyond description and the country, colonial, and London troops expressed their opinion equally emphatically in their own peculiar way. Think again of the need of systematic wind observation along the whole front of attack, the disorganisation and "gas alert" conditions imposed on the favourable night, the possibility of postponement, and we can only draw one conclusion. There must have been some imperative need or justification of cloud gas attack for the army to have encouraged or even tolerated its continuance. There is no difficulty in understanding why gas attack was so exceedingly unpopular among the staffs in the early stages of the war. Later, however, when they realised the enemy casualties that were being created by the gas, and what a large part it was taking in the war of attrition, the opposition and lack of appreciation vanished. Further, when the projector arrived to produce similar effects with less demand upon infantry personnel, and less dependence on the wind, the whole tone of the army towards gas was changed, and it became almost popular.

The Riddle of the Rhine

The peculiarity of cloud gas attack was the concentration of all this effort of preparation within a few days. In terms of military efficiency, the amount of energy expended was fully justified by the casualties produced. We know that some of our cloud attacks were responsible on one night for many thousands of casualties, and the amount of artillery effort to give such a result would probably have been considerably larger. But under normal conditions of warfare, such artillery effort would have been expended over a much longer period of time.

The Livens Projector.—The Somme offensive witnessed the use of a new British gas weapon which became of the utmost importance. This was the mortar known as the Livens Projector. Its origin dates back many months, however, and is of considerable interest. A British engineer, Lt. Livens (afterwards Major, D.S.O., M.C.) of the Signal Corps, was inspired to constructive and aggressive thought on the gas question by a double motive. He quickly realised the tactical weakness of the German method at Ypres, once shorn of its vast initial possibilities of surprise. He saw the advantage of being able to command the point or locality of incidence of the cloud, instead of being limited to the actual trench front. Prompted by a direct personal interest in the huge loss sustained by the *Lusitania* outrage, he determined to find a practical outlet for his feelings by developing his views on the future of gas clouds. In a few months the general principles of the projector were defined and a crude specimen resulted. Caught up, however, in the gas organisation, preparations for the cloud attack at Loos absorbed all his attention and energies and the consequent reorganisation found him developing a flammenwerfer and training a company for its use.

THE LIVENS PROJECTOR—II.

A working party fitting electric leads and adjusting bombs prior to discharge. This work occurs at night.

To face page 61

It was really the Somme battle which gave him the first opportunity to carry his idea into offensive practice. This arose in front of High Wood, which was a veritable nest of German machine gunners in such a critical tactical position as to hold up our advance in that region. The huge stationary flammenwerfer had recently been used by Major Livens and his company against a strong point in front of Carnoy in the assault of July 1st. Here again the effect of flame was limited even more than that of cloud gas by dependence on a fixed emplacement. It was quickly grasped that the solution was to be found in the application of the projector principle to the use of oil for flame and a crude projector was devised for the emergency, using oil cans as mortars, burying them in the earth for two-thirds of their length and employing water cans as bombs.

As soon as the possibilities of the weapon were seen its development was pressed. The usual Livens Projector consisted of a simple tube mortar or projector closed at one end, and fitted with a charge box on which rested the projectile. By an electrical arrangement and suitable communications, large numbers, sometimes thousands, of these projectors could be discharged at a given moment. In this way quantities of gas, comparable with the huge tonnages employed in the normal stationary cloud attack, could be used to produce a cloud which would originate, as cloud, as far as a mile away from the point of discharge. In other words, the advantages of cloud attack could be used with a much smaller dependence on wind direction, and with a much greater factor of local surprise. Thus when the partially perfected and efficient weapon was used in large quantities during the British Arras offensive in April, 1917, the German Army was thrown into great consternation.

But for the fact that protection had developed so strongly on both sides, the use of the Livens Projector would have gone far towards a decision.

The simplest way to illustrate the peculiar value of the projector will be to quote from one or two of the many Intelligence reports collected. Thus from a captured document dated July, 1917, belonging to the 111th German Division, signed Von Busse, we have: "The enemy has combined in this new process the advantages of gas clouds and gas shells. The density is equal to that of gas clouds, and the surprise effect of shell fire is also obtained. For the bombardment the latter part of the night is generally chosen, in a calm or light wind (the direction of the latter is immaterial). The enemy aims essentially at surprise. Our losses have been serious up to now, as he has succeeded, in the majority of cases, in surprising us, and masks have often been put on too late. . . . As soon as a loud report like a mine is heard 1000-1500 metres away, give the gas alarm. It does not matter if several false alarms are given. Masks must not be taken off without orders from an officer. Men affected, even if apparently only slightly, must be treated as serious cases, laid flat, kept still, and taken back as soon as possible for medical treatment. Anti-gas officers and Company Commanders will go through a fresh course of training on the above principles." The influence of gas discipline is borne out by another captured statement that they could only attempt to "reduce their losses to a minimum by the strictest gas discipline." Again, from a prisoner we learn that "every time a battalion goes into rest, masks are inspected and a lecture is delivered by the gas officer on British gas projectors, which are stated to be the most deadly form of warfare." So great was the impression formed by the introduction of the projector that uneasiness at the front was reflected later on in the Press. Thus, quoting from reference to the military discussion before the main committee of the Reichstag. "Casualties from enemy poison gas admit on the whole of a favourable judgment, as the harm involved is only temporary, and in most cases no ill after-effects persist" (*Tägliche Rundschau*, 24.4.18). "Cases of gas poisoning are not as a rule accompanied by harmful consequences, even though the treatment extends sometimes over a long period" (*Vorwärts*, 25.4.18). Based on the later mustard gas casualties these statements would have been more truthful. As it was, they afforded poor consolation to the German people.

British Gas Shell.—The British first used shell gas as lachrymators, in trench mortar bombs, in small quantities, during the battle of the Somme, but for the first time, during the battle of Arras, 1917, our supplies of gas for shell were sufficient for extensive and effective use. Our success can be measured by the report dated April 11th, 1917, from the General Commanding the first German Army, on "Experiences in the Battle of Arras," in which he says: "The enemy made extensive use of gas ammunition against our front positions as well as against batteries." "The fighting resistance of the men suffered considerably from wearing the mask for many hours." Artillery activity seems to have been paralysed by the effects of the gas.

In a general comparison of British and German methods of gas warfare,[1] General Hartley tells us "our methods improved rapidly during 1917. At first we neglected, almost entirely, the question of rate of firing, but we soon arrived at the method of

[1] Journal of the Royal Artillery, February, 1920.

crashes of lethal shell. These got the surprise concentrations of gas which proved so effective, and we realised that the number of shells required to produce an effect was much bigger than we thought originally. At Messines gas was used in much the same way as at Arras."

German Gas Shell Development, 1916.—The main evidence of Allied reaction was to be found in the intensive development of cloud gas attacks, but during the same period the Germans, who appeared to be abandoning the use of cloud gas, were making steady efforts to regain their initiative by the comprehensive development of shell gas. Thus, to quote from General Hartley's report to the British Association, "In the Summer of 1916 chlor-methyl-chloro-formate with toxic properties similar to those of phosgene was used against us in large quantities during the battle of the Somme. Later this was replaced by tri-chlor-methyl-chloro-formate, a similar liquid, which was used until the end of the war as the well-known Green Cross shell filling. The use of phosgene in trench mortar bombs also began in 1916." Many of those on the front in 1916 will remember the surprise gas shell attack of December of that year, on the Baudimont gate at Arras. We were fortunately let off lightly with little over 100 casualties, but the effect was to tighten up gas discipline all along the line. The appearance of the new substances represented definite German progress and had definite military results, but they lost decisive value owing to the relative inefficiency of German gas shell tactics.

Consideration of the Allied reaction must include some reference to the appearance of the American Army in the field. The Americans during their more or less educational period gave serious attention to the gas question, and showed almost immediately, by their preparations, that they attached enormous importance to the new weapon.

Main Features of the Period.—It is difficult to generalise. But the following features appear to characterise the period under discussion. In the first place we see German policy tending towards the use of gas projectiles containing a variety of organic substances. Secondly, we have the British exploitation of cloud gas attack both in magnitude and method. The Livens Projector provides the third important feature. Fourthly, we note the somewhat tardy development of the British use of gas shell. A number of causes, no doubt, unite in responsibility for the above. But whether due to definitely framed policy on our part, or merely to the hard facts of the case, one important factor seems largely responsible. It is the relative ease of production by Germany as compared with ourselves. When German military opinion tended towards the development of gas shell, a variety of substances came quickly to hand, not only from German research sources, but in quantity from the dye factories. No such quick response could have met, or actually did meet, the demands of Allied military policy. Whatever ideas emanated from our research organisations, there was no quick means of converting them into German casualties. It is true that we could obtain chlorine and later phosgene in bulk and devote them to the exploitation of the older gas appliances in cloud methods. But British chemical supply was weak, owing to the absence of a strong organic chemical industry. In other words, German flexibility of supply meant flexibility in meeting the requirements of military policy, and, given sound military policy, this flexibility meant surprise, the essence of successful war.

CHAPTER IV

The chemical struggle became very intense in the Summer and Autumn of 1917. Projector attacks multiplied, the use of chemical shell increased on both sides, allied and enemy gas discipline was tightened up, officers and men acquired a kind of gas sense, a peculiar alertness towards gas. The home front was strengthened in England and France by reinforced and sounder organisations, and by the vigorous steps taken by America. The Germans began to reap the benefit of their gas shell policy. At the end of 1916, as a result of a review of the production situation, they had arrived at the so-called Hindenburg Programme. This included a large output of gas for shell, and from its realisation the Germans acquired a momentum which kept them ahead well into 1918. It is a very clear indication of the progress made by Germany in research, that the sudden expansion in manufacture required by the Hindenburg Programme found a number of new efficient war chemicals ready for production.

The Mustard Gas Surprise.—The next big surprise came from Germany. Units in the line at Nieuport and Ypres in July, 1917, were the first to experience it. Some were sprinkled and some deluged with a new type of German shell chemical which, in many cases, evaded the British gas discipline, and mustard gas, unrecognised, caused many serious cas-
66

ualties. Even those who wore the mask were attacked by the vesicant or blistering influence of the gas. The matter is vividly expressed in a letter, given below, which I received from an officer wounded in the Nieuport attack:

"I was gassed by dichlor-diethyl sulphide, commonly known as mustard stuff, on July 22nd. I was digging in (Livens Projectors), to fire on Lambartzyde. Going up we met a terrible strafe of H.E. and gas shells in Nieuport. When things quietened a little I went up with the three G.S. wagons, all that were left, and the carrying parties. I must say that the gas was clearly visible and had exactly the same smell as horseradish. It had *no* immediate effect on the eyes or throat. I suspected a delayed action and my party all put their masks on.

"On arriving at the emplacement we met a very thick cloud of the same stuff drifting from the front line system. As it seemed to have no effect on the eyes I gave orders for all to put on their mouthpieces and noseclips so as to breathe none of the stuff, and we carried on.

"Coming back we met another terrific gas shell attack on Nieuport. Next morning, myself, and all the eighty men we had up there were absolutely blind. The horrid stuff had a delayed action on the eyes, causing temporary blindness about seven hours afterwards. About 3000 were affected. One or two of our party never recovered their sight and died. The casualty clearing stations were crowded. On August 3rd, with my eyes still very bloodshot and weak and wearing blue glasses, I came home, and went into Millbank Hospital on August 15th."

These early mustard gas attacks caused serious gaps amongst the troops assembling for the Northern offensives. The gas was distinctly a new departure.

Effective in low concentrations, with very little odour, and no immediate sign of discomfort or danger, very persistent, remaining on the ground for days, it caused huge casualties. Fortunately, its most fatal effects could be prevented by wearing a respirator, and only a very small proportion of mustard gas casualties were fatal.

The insidious nature of the gas and the way in which it evaded the gas discipline is shown in the following example from an official report: "A battery was bombarded by the new gas shell from 10 p.m. to 12 midnight and from 1.30 to 3.30 on the night of 23rd-24th July. The shelling then ceased and at 6 a.m., when the battery had to carry out a shoot, the Battery Commander considered the air free from gas, and Box Respirators were accordingly removed. Shortly afterwards several men went sick from gas poisoning, including the Battery Commander. On previous nights they had been fired at with gas shell in the same way, but found it safe to remove Box Respirators after a couple of hours. On the occasion in question the air was very still and damp." In another case an officer in the Boesinghe sector, during the gas bombardment on the night of the 22-23 July, adjusted the mouthpiece and nose-clip, but left the eyes uncovered. His eyes were seriously affected, but he had no lung symptoms on the morning of the 24th.

Mustard gas (or Yellow Cross, as it was called officially by the Germans) was the war gas *par excellence* for the purpose of causing casualties. Indeed, it produced nearly eight times more Allied casualties than all the various other kinds of German gas. It was used for preparation a considerable time before the attack, or during the attack, on localities and objects with which the attackers would have no contact.

Blue Cross.—Another new type, the German Blue Cross, was introduced about the same time. This represented at different times diphenylchlorarsine, diphenylcyanarsine and other arsenic compounds. The Blue Cross compound was contained in a shell with high explosive. The enemy expected that the shell burst would create such a fine diffusion of the compound that it would penetrate our respirator mechanically, and then exercise its effects. These, violent irritation of the nose and throat, nausea and intense pain, would cause the removal of the respirator and allow other lethal gases to have full play. Fortunately, the German hopes of penetration were not realised, but they were, no doubt, continuing to develop the vast possibilities of the new method.

German Emphasis on Gas Shell.—The Green Cross or lethal filling was another type of German gas shell. Green Cross covered such compounds as phosgene and chlor-methyl chloroformate. Although these caused fewer casualties than mustard gas, they were relatively more fatal. Schwarte's book tells us that, "After the introduction of the Green Cross shell in the summer of 1916, at Verdun over 100,000 gas shell were used to a single bombardment."

From the time of the first use of mustard gas until the terrific gas shell attack of March, 1918, the Germans persistently used their new types against us with considerable effect. Even when the period of surprise effect with mustard gas was over, the number of casualties caused by it was considerably greater than during the months when the Germans were firing only non-persistent lethal shell of the Green Cross type. The Germans regarded these

69

shell gas developments as largely responsible for our failure to break through in the Autumn of 1917.

The German Projector.—During this period they also developed a projector. Their first use of it was again co-ordinated with an attempt at surprise. Fortunately, protection and gas discipline had reached such an efficient state that normal "alert" conditions of the front line system were largely able to counter the use of this new device by Germany. The first attack was against the French at Rechicourt on the night of December 5th-6th.

On the night of December 10th-11th, 1917, they fired several hundred projectiles on the Cambrai and Givenchy sectors of the British line. In both cases the gas bombs were fired almost simultaneously into a small area including our front and support lines. The bombs appeared to have been fired from the enemy support line, as observers state that they saw a sheet of flame run along this line, followed by a loud explosion. The bombs, which emitted a trail of sparks, were seen in the air in large numbers and made a loud whirring noise. They burst with a large detonation, producing a thick, white cloud. The discharge was followed immediately by a bombardment with H.E. shrapnel and gas shell, and a raid was attempted south of Givenchy. We learn that so strong was the gas discipline that in many cases respirators were adjusted before the arrival of the bombs, the resemblance to our projector attacks having been established at once. When this was done practically no casualties occurred. Again, to show the efficiency of British protection against projector gas, we learn from official reports that, "At one point five bombs burst in a trench without harming the occupants. It should be remembered that the British box respirator protects against very high concentrations of gas which pass at once through the German mask." Similar discharges were made against the French on two occasions in December, and against the Lens sector on December 30th. The compounds used in the bombs were phosgene and a mixture of phosgene and chlorpicrin. These attacks increased in number during the ensuing months.

German Projector Improvements.—The Germans developed a longer range modification and would undoubtedly have exploited this weapon very considerably but for the trend of the campaign. The Allied advance in 1918 uncovered a number of enemy dumps. Amongst the most interesting was one which contained a number of a new type of projector.

A prisoner of the 37th pioneer gas battalion, captured on August 26th, had said that they were to practise with a new type of projector with a range of 3 kilometres, the increased range being obtained by rifling the bore of the projector. He stated that the intention was to use the longer range weapons in conjunction with the old short range projector, using the new type to deal with the reserve positions. The capture of the dumps referred to above revealed the truth of his statement. Two kinds of bombs were used, one containing H.E. and the other small pumice granules impregnated with phosgene. This was an ingenious attempt to produce a persistent but highly lethal gas by physical means, for hitherto the highly lethal gases had only been slightly persistent. The new projector had a calibre of 158 mm. and was termed the "Gaswerfer, 1918." The importance of this new projector cannot be overestimated. Its large scale use would, undoubtedly, have resulted in imposing stringent gas alert conditions at greater distances from the front line.

71

Dyes in Gas Shell.—Another interesting German development of this period was the use of certain dyes or stains in gas shell. After gas bombardments in the winter of 1916-17, the snow was seen to be covered with coloured patches. These coincided with the bursts of the shell. Analysis of the earth showed that the colour was due to the presence of an actual dyestuff. A number of explanations were advanced to account for the use of the colour, of which the most probable claimed its employment for the identification of affected localities several hours or even days after the bombardment. This was especially the case with persistent types. As the explosive charge of chemical shell was feeble, some such means of identification was necessary. It may be that the Germans expected that troops advancing after such bombardments would be helped by the splashes of colour, and that these earlier attempts were purely experimental.

German Flame Projectors.—We have already referred to the use of flame projectors by the enemy, and a picturesque account of their development and use in the later stages of the campaign is found in an extract from the *Hamburger Nachrichten* of the 9th of June, 1918:

Their Origin.—"Our Flammenwerfer troops owe their origin to a mere incident. Their present commander, Major R., when an officer of the Reserve, received the order, during peace manœuvres, to hold a certain fort at all costs. During the sham fight, having employed all means at his disposal, he finally alarmed the fire brigade unit, which was under his orders as commander of the fort, and directed the water jets on the attacking force. Afterwards, during the criticism of operations in the presence of the Kaiser, he claimed that he had subjected the attackers to streams of burning oil. The Kaiser thereupon inquired whether such a thing would be possible, and he received an answer in the affirmative.

"Long series of experiments were necessary before Engineer L. succeeded in producing a combination of various oils, which mixture is projected as a flame on the enemy by means of present day Flammenwerfer.

"Major R. occupied himself in peace time with fighting fire as commander of the Munich Fire Brigade. The 'Prince of Hades,' as he is called by his 'fire spouters,' enjoys great popularity among his men as well as among the troops to whose assistance he may be called. He can look back on an important development of his units. Whereas in January, 1915, Flammenwerfer troops consisted of a group of 36 men, to-day they constitute a formation with special assault and bombing detachments, and are furnished with all requisites for independent action. In reading Army Communiques, we often find mention of these troops. If difficulty is experienced in clearing up an English or French Infantry nest, the 'Prince of Hades' appears with his hosts and smokes the enemy out. That conditions of membership of this unit hardly constitute a life insurance policy is obvious; nor is every man suitable. Special men who are physically adapted and who have given proof of keenness in assault are necessary for such work."

Further Flame Development.—Specimens of a very neat portable German Flammenwerfer were captured in August, 1917. It contained three essential parts: a ring-shaped oil container surrounding a spherical vessel containing compressed nitrogen, which was used to expel the oil, and a flexible tube of rubber and canvas carrying the jet. The whole

73

was arranged to be carried on the back. At about this time prisoners stated that men were transferred to the Flammenwerfer companies as a form of punishment.

The Germans were fond of using the Flammenwerfer during counter-attacks and raids in which the morale factor is so important. Thus in September, 1918, in a raid against the British during our great offensive, the German raiding party was heralded by a shower of stick bombs and the Flammenwerfer men followed. The bombing party advanced under cover of these men, the smoke from the flame throwers acting as a screen. British experience was that the calm use of machine-gun fire soon put German flame throwers out of action, and it is clear that the Germans themselves realised this weakness of isolated flame attacks for, in one of their documents issued by German G.H.Q. in April, 1918, they said: "Flammenwerfer have been usefully employed in combats against villages. They must be engaged in great numbers and must fight in close liaison with the infantry, which helps them with the fire of its machine-guns and its grenades."

The 1918 Offensive.—Some idea of the importance of these developments and of the scale on which they were exploited in the later campaigns of the war can be obtained by briefly examining the German plans for the use of gas in their 1918 offensive, and their execution: *Die Technik im Weltkriege* tells us: "During the big German attacks in 1918, gas was used against artillery and infantry in quantities which had never been seen before, and even in open warfare the troops were soon asking for gas."

The Yellow and Blue Cross shells first introduced into operation in July, 1917, were not incorporated into comprehensive offensives until March, 1918. Owing to the exigencies of the campaign, the initial surprise value of these gases was subordinated to the later large scale use in the great offensive. In December, 1917, the German Army was instructed anew regarding the use of the new gas shell types for different military purposes, laying great stress on the use of non-persistent gas for the attack. Fortunately for us, the gas shells destined for this purpose were not relatively so efficient as the German persistent types, which were devoted to the more remote preparation for attack and to defensive purposes. Their penetrating Blue Cross types were a comparative failure. Although plans emphasised the importance of this gas for the attack, facts later gave greater prominence to the use of the persistent Yellow Cross shell for defensive purposes in the great German retreat.

Ludendorff's Testimony.—Ludendorff, himself, emphasised the great importance which was attached to gas in this offensive. He says[1]: "And yet our artillery relied on gas for its effect, and that was dependent on the direction and strength of the wind. I had to rely on the forecast submitted to me at 11 a.m. by my meteorologist, Lieutenant Dr. Schmaus. Up till the morning of the 20th strength and direction were by no means very favourable; indeed, it seemed almost necessary to put off the attack. It would have been very hard to do. So I was very anxious to see what sort of report I should get. It was not strikingly favourable, but it did indicate that the attack was possible. At 12 noon the Army Groups were told that the programme would be carried out. Now it could no longer be stopped. Everything must run its course. G.H.Q.

[1] *My War Memories.* Hutchinson & Co., 1919.

higher commanders and troops had all done their duty. The rest was in the hands of fate, unfavourable wind diminished the effectiveness of the gas, fog retarded our movements and prevented our superior training and leadership from reaping its full reward."

Preparations for Assault;—Gas Defensive at Armentieres.—For twelve days prior to their March assault the Germans used mustard gas over certain areas, and the non-persistent types for other localities. As an example of the first method, we can state that nearly 200,000 rounds of Yellow Cross shell were used on the 9th March, and caused us heavy casualties. The actual attack at once confirmed our suspicions of enemy intention to break through on the territories which were not infected by the persistent mustard gas. In the second case, of the non-persistent types of Blue and Green Cross, bombardments of tremendous intensity occurred for several hours before the assault, on all defensive positions and organisations for several miles behind the front line. Millions of rounds must have been used. Although not without serious effect on the campaign, this furious gas attack did not fully justify expectations. The failure of mask penetration by the Blue Cross shell prevented the full possibilities of Green Cross coming into play. To illustrate the specific use of gas in this great offensive, and the organic way in which it was co-ordinated in the plan of attack, we quote from a recent statement by General Hartley.[1] Referring to the gas shelling immediately before the extension of the attack to the north of Lens on 9th April, he explains, "Between the 7th April and 9th April there was no gas shelling between the La Bassee Canal and Armentières,

while there was heavy Yellow Cross shelling immediately south of the Canal, and Armentières had such
a heavy bombardment that the gutters were running
with mustard gas. This indicated the probability
of an attack on the front held by the Portuguese,
which occurred on 9th April, Blue and Green Cross
being used in the preliminary bombardment." The
Portuguese front lay between the two Yellow Cross
regions.

Fixed Gas Barrage at Kemmel.—Another most
interesting example is also quoted, dealing with the
shelling preceding the attack on Kemmel on 25th
April. "This is an interesting case, as non-persistent
Blue Cross shell were used within the objective and
Yellow Cross just behind it, indicating that on 25th
April the enemy did not intend to go beyond the
line they gained."

Percentage of Chemical Shell.—Some idea of the
importance which the Germans attached to their
chemical ammunition, as distinct from explosives,
can be gathered from the following extract from a
captured order of the Seventh German Army, dated
May 8th, 1918, giving the proportion of chemical
shell to be used in the artillery preparation for the
attack on the Aisne on 27th May, 1918.

"(*a*) Counter-battery and long range bombardments.
> For 7.7 c/m field guns, 10.5 c/m and 15 c/m
> howitzers and 10 c/m guns; Blue Cross 70%,
> Green Cross, 10%; H.E. 20%, long 15
> c/m guns fire only H.E.

(*b*) Bombardment of infantry positions.
> (i) Creeping Barrage.
> For 7.7 c/m field guns, 10.5 c/m and 15
> c/m howitzers; Blue Cross 30%, Green

77

Cross 10%, H.E. 60%, 21 c/m howitzers fire only H.E.

(ii) Box Barrage.

For 7.7 c/m field guns, 10.5 c/m howitzers and 10 c/m guns; Blue Cross 60%, Green Cross 10%, H.E. 30%."

What more striking demonstration is needed than these extraordinarily high percentages?

Gas Retreat Tactics;—General Hartley's Analysis.—No Yellow Cross shell were to be used in the bombardment, but, as mentioned above, there was a complete change of tactics in their retreat, during which they attempted to create a series of barriers by literally flooding areas with mustard gas. This defensive use of mustard gas was most important. Again, quoting General Hartley, "Yellow Cross shell were used much farther forward than previously, bombardments of the front line system and of forward posts were frequent, and possible assembly positions were also shelled with this gas. On more than one occasion when an attack was expected the enemy attempted to create an impassable zone in front of our forward positions by means of mustard gas. Their gas bombardments usually occurred on fronts where they had reason to fear an attack, with the idea of inflicting casualties in areas where troops might be massing. It was instructive to note how supplies of Yellow Cross shell were switched from the Third to the First Army front late in August when they became nervous about the latter sector. In Yellow Cross they had an extremely fine defensive weapon, which they did not use to the best advantage, for instance, they neglected its use on roads and did not hamper our communications nearly as much as they might have done. As our offensive progressed their gas shelling became less organised, and one saw very clearly the superior value of a big gas bombardment as compared with a number of small ones. In the latter case it was usually possible to evacuate the contaminated ground and take up alternative positions, while in the case of a bombardment of a large area such as the Cambrai salient, the difficulty of doing so was greatly increased, and consequently casualties were higher. During our offensive it was not possible to exercise the same precautions against gas as during stationary warfare, and the casualties were increased on this account."

Percentage of German Gas Shell in Enemy Dumps.—A test of the importance attached by any army to the different types of ammunition which it uses can be made by examining the percentage of such types of shell in a number of ammunition dumps assembled behind the front line for some specific operation, or part of a campaign. An examination of German production from this point of view is very interesting, and also brings out a significant point. The normal establishment of a German divisional ammunition dump in July, 1918, contained about 50 per cent. of gas shell. The dumps captured later in the year contained from 30 per cent. to 40 per cent. These figures are significant, for they show how much importance the German Army attached to gas shell. When we think of the millions of shell and of the huge quantities of explosives turned out by our own factories to fill them, and when we realise that for a large number of gun calibres the Germans used as many shell filled with gas as with explosive, some idea of the importance of gas in the recent war and of its future possibilities can be obtained. Further, when we realise that the production of ex-

79

plosives can be controlled and inspected during peace, but that no such control can exist for chemical warfare products, the significance for the future stands revealed.

Forced Exhaustion of Stocks.—It might be thought that the lower percentages found later on in the year were an indication of the decreasing importance of chemical shell. Examining the case less superficially, however, we soon see that this lower percentage has an entirely different meaning. In the first place, we know that the German factories were still pressing on to their maximum output at the time of the Armistice. New units were being brought into operation. Secondly, we have seen how huge quantities of mustard gas were diverted to those particular German armies which were most threatened by the final Allied offensive, indicating that certain portions of the German front were being starved for chemical shell. The truth of the matter is that the Germans had accumulated enormous stocks for their great offensive and that they had expended these stocks at a greater rate than their factories could replace them. We learn from Schwarte's book that, "Although the production of Yellow Cross almost reached 1000 tons a month, yet finally the possibilities of use and the amount required were so great that only a much increased monthly output would have been sufficient."

Yperite, French Mustard Gas.—During this period the volume of allied gas activities also increased considerably. But until June, 1918, our success was due to the development of more successful tactical methods rather than to any specific chemical surprise.

Very great credit is due to the French for having

produced large quantities of mustard gas by the above date.

Judging from the German Intelligence Reports the surprise effect of the French production was almost as great as that obtained by the earlier German use. It again evaded the gas discipline of the troops, and we find the German staff laying enormous emphasis on this question, which was already very prominent in their general and operation orders. The occasion provided a very striking example of German belief in their absolute predominance in production. They were largely justified in this belief, but it carried them too far. They explained the use of mustard gas by the French as due to the use of German mustard gas obtained from "blind" German shell!

Effect on German Gas Discipline.—British mustard gas was not in use in the field until September, 1918, but the French was a great success, and probably contributed to no small extent to the final allied success in the 1918 campaign. The French termed mustard gas "Yperite" after Ypres, the place where it was first used. As far as such terms can be applied to any weapon, Yperite arrived to spread panic and terror amongst the German formations. A document captured by the Sixth French Army shows that Yperite used on the 13th June against the 11th Bavarian Division was the chief cause of the precipitate retreat of this Division. The Seventh German Army refers to another bombardment on the 9th of June, in which the casualties exceeded five hundred.

It is curious to note that although the Germans had so preached the superiority of their gases and gas methods, serious blows by the Allies found the German gas discipline unequal to them. It is no exaggeration to say that the use of mustard gas by the French, and later by the British, and the British projector, produced, on each occasion, in the German ranks feelings allied to panic. This is reflected in the many orders which have been captured from army and other headquarters enforcing and even appealing for gas discipline amongst the troops. Thus, almost immediately after the first French use, Ludendorff, chief of the German General Staff, issued a special detailed order on the subject, and the German document captured by the French can be taken as representative. "Our Yellow Cross has caused much damage to the enemy, formerly less protected than now. But as a natural sequence he had developed through it a gas discipline which can certainly be taken as model. On this account enemy troops have been able to cross, at once and without loss, areas which their artillery had just bombarded with gas. We also must train our troops to an excellent standard of gas discipline if we expect to avoid the grave dangers which threaten the fighting forces of our army." By the time of the Armistice France had produced nearly 2000 tons of mustard gas, British and American production was rapidly increasing, so that the output was attaining stupendous proportions. Some idea of the importance of chemical warfare in the campaigns of 1917-1918 can be obtained from the following figures:

Allied Gas Statistics.—Between November, 1917, and November, 1918, France produced more than five millions of her latest type of respirator. The British figure was probably higher. From April to November, 1918, the French filled nearly two and a half million shell with mustard gas. From the 1st of July, 1915, to the latter date more than seventeen million gas shell were completed by the French.

In addition to these huge gas shell figures we must remember the chemical operations from projectors and as cloud gas. During the period the British averaged fifty large scale operations of this type per month, sometimes discharging monthly three hundred tons of gas. The total French production of chlorine and poison gas for chemical warfare approached 50,000 tons, a large proportion of which production occurred during 1917 and 1918. The British was of the same order, but German production was at least more than twice as high, showing what great use they made of gas shell. The huge American programme might have reduced the margin, but no limits can be placed on German possibilities and elasticity in production.

Critical Importance of Rapid German Production.—These figures are misleading inasmuch as they give no indication whatever of the relative difficulties and corresponding rapidity of action on both sides. As a general rule, where the German lag between the approval of a substance and its use in the field covered weeks, our lag covered months. Owing to efficient production, chemical warfare was an infinitely more flexible weapon in German hands than in ours. This will be readily understood when we analyse, later, the methods of production of some of the chief German war gases. In general, German development of these complicated substances provided a series of examples of the ease and rapidity of production of organic substances by the dye industry. On the other hand, except in very few exceptional cases, British and French production, although we cast no reflection on the energy or skill of any concerned, was exceedingly slow and costly by comparison. The Germans used mustard gas in July, 1917. We identified it a few days afterwards. But the first fruits of

83

allied production were not in the field for eleven months. British material was not used until a month or two before the Armistice. Further, in this case, we were convinced of the value of the substance almost from the first day of its use by the enemy. We will endeavour to throw light upon this in our review of production.

The period of intensive chemical warfare may be regarded as the proof of the German experiment of 1915-1916. Shed of their trial nature, the chemical weapons played a logical and increasingly dominating part in the campaign. They were surely destined to play a much more prominent part had the period of stabilised warfare continued. Projector cloud gas would have assumed greater importance as a casualty producer. But we will leave such considerations for a future chapter.

CHAPTER V

We have no desire nor intention to give a detailed historical account of the above. The ramifications of Allied organisations were so numerous, the number of persons concerned so great, the sacrifices made so heavy, that only an exceedingly lengthy account could hope to do justice to individuals. In addition, such an account would not serve our purpose. We wish to show, as briefly as possible, how the different Allied organisations were bound up in an organic way with the campaign, how they compared with those of the enemy, and what lesson the comparison may contain for the future.

Two facts stand out in such a comparison. We are struck with the extreme simplicity of the German organisations, as we know them, and the great complexity and multiplicity of the Allied departments as we saw them. We must admit from the beginning that we know least of the German home organisations for research and production, but our knowledge is sufficient to reveal their simplicity. The Inter-Allied Commission of Control may, and certainly should, obtain full information, but at present the matter stands as follows.

German Research.—The Germans relied upon two main and very strong centres for research. They have already been indicated as the Kaiser Wilhelm Institute, under the direction of Professor Haber,

and the enormous research organisations of the I.G.
There are various references to internal gas organisation in captured documents. It appears that they
received their final form late in 1917. A great gas
school (Heeres-Gaschule) was instituted in Berlin
where there were also central depots for anti-gas
inspection and material. Rather earlier than this the
Kaiser Wilhelm Institute was definitely appointed as
the official research centre. The War Ministry had
a chemical section named A.10, which dealt with gas
questions. It is rumoured, and there is strong reason
to believe, that the I.G. was largely staffed by officers
of the Reserve before the war. Whatever their pre-war
associations, if any, with the War Ministry, hostilities
must have found them keenly alive to the possibilities
of their unique research and organic chemical producing facilities. It is inconceivable that this military
personnel should not have greatly assisted the I.G. in
its operations, inventions and general assistance for
the army.

It appears that the subdivision of work left the
direction of chemical research in Berlin, possibly at
the above Institute, while the bulk of the work of
preparing the new compounds, and developing manufacturing processes for approved substances, occurred
in the laboratories of the I.G.

Leverkusen.—We know, for example, that a very
large number of substances was produced at Leverkusen and samples forwarded to Berlin, of which only
a few were finally approved for production. The
physiological work and field tests were certainly associated with the Berlin organisation, but it is not
clear how much of this work occurred within the I.G.
An Allied mission to Leverkusen reported as follows :—"It was emphatically stated that no means of
testing the products were resorted to beyond inhala-

tion and testing the effect of the substances on the staff, but this statement must be accepted with reserve." This is particularly so as we know that large numbers of respirator-drums had been made in this factory, and that a gas school existed at Leverkusen in 1915.

A member of another Allied mission was informed by one of the staff at Leverkusen that the authorities there were well aware of the difficulties in chemical warfare, apart from production, for they had some experience in the designing and testing of chemical shell. It may be that the German Government relied upon the I.G. for such work in the early stages of the chemical war, pending the development of official organisation. When we remember, however, that at Leverkusen alone there was a staff of 1500 technical and commercial specialists, apart from thousands of workpeople, before the war, and that the latter were increased by 1500 during the war, we find it difficult to place a limit on the services which might have been rendered by this research centre alone. The opinion of the members of the Hartley Commission[1] was, that much thought and attention had been given to chemical warfare by the chemists of the Company.

Höchst.—A great volume of chemical warfare research occurred also at Höchst. "The admission was made that the research department of the factory was continuously employed during the war on the preparation of substances suitable for chemical warfare, many hundreds being prepared and sent to Berlin for examination. The firm employed 300 academically trained chemists in peace time, but during the war many more were engaged, partly for

[1] A post-armistice inter-allied mission of experts, to the Rhine chemical factories, March, 1919.

research and partly because all shell filling was carried out under the supervision of trained chemists."

Ludwigshafen.—The most influential branch of the I.G. was, undoubtedly, the Badische Anilin und Soda Fabrik. It might have been expected, as they shared largely in production, that a considerable amount of chemical warfare research would occur at these works, but this was emphatically denied to Allied missions. It may be, however, that as the nitrogen fixation enterprise was developed there, requiring a large amount of technical development and control, this was considered a sufficient contribution to the general cause.

Early Formulation of Policy.—In examining what signs we have of the organisation and policy underlying chemical warfare research and production in Germany, we are struck by the fact that all the substances used with such dire effect against us during the war must have been approved for production by the Government at a relatively early date. The following table, assembled from information supplied by the German factories, brings this point out very clearly.

War Chemical.	Factory.	Production Began.	First Use in the Field.
Diphosgene (Green Cross)	Höchst Leverkusen	Sept., 1916 June, 1915	Summer, 1915
Mustard Gas (Yellow Cross)	Leverkusen	Spring, 1917	July, 1917
Diphenyl-chlorarsine (Blue Cross)	Höchst	May, 1917	July, 1917
Diphenyl-cyanarsine (Blue Cross)	A.G.F.A.?	Feb., 1918	June, 1918
Ethyl-dichlor-arsine (Blue Cross)	Höchst	Aug., 1917	March, 1918

We have chosen the later products to establish the point, for it is self-evident for the earlier products, some of which were made before the war.

Movements of Personnel.—The movements of German chemical personnel give us a clue as to the main tendencies in their chemical warfare policy. The factories were called upon to produce, as we have already shown, towards the end of 1914, but this production largely involved the use of substances already manufactured on a certain scale. Large scale production of the more advanced types of war chemical seems to have been directly stimulated by the Hindenburg programme, in connection with which the Companies withdrew large numbers of their skilled workers from the front.

German Simplicity of Organisation.—We can safely conclude from the above that Germany required no cumbersome government mechanism for the preparation of new war chemicals, for the semi-industrial work in developing processes for approved substances, nor for their production. By relying on the I.G., the Kaiser Wilhelm Institute, and probably some other organisation for field and physiological tests, Germany escaped the necessity for comprehensive government organisation, the development of which was such a handicap to Allied countries. It is certainly very suggestive that we only met, in the field, substances approved before the summer of 1917. It is with great interest and a certain amount of apprehension that we speculate upon the research developments after that period with which the war did not make us immediately acquainted. If this early period produced such effective results as mustard gas, Blue Cross compounds, and the different cloud substances, what hidden surprises were matured in the later period? This feature of simplicity, of linking up a new war with an old peace, activity was paralleled somewhat in the field organisation. We have seen how Germany created special formations for cloud attacks, but for a time practically abandoned them, throwing most of her chemical warfare production into shell. In other words, she substituted a normal weapon, the artillery. We, on the other hand, largely impelled by the enforced simplicity of our production, tended more towards the development of special formations and special weapons for cloud production, but with such success that the German Pioneer formations, after being practically dropped, found a use in developing and using our new weapon, the Livens Projector.

German Organisation at the Front;—The Gas Regiment.—It is probable that the earliest form of German organisation at the front consisted in the liaison between Professor Haber and the German G.H.Q. It will be remembered that Ludendorff, discussing cloud and shell gas, refers to this co-operation, stating:[1] "Geheimrat Haber proved of valuable service in this connection with the use of gas." It was also rumoured soon after the first German attack that the organisation and preparation of the latter were under the scientific guidance of this renowned Professor. The attack was carried out by the 35th and 36th Pioneer Regiments, each furnished with chemically trained officers who were specially detailed for gas warfare.

The importance of protection was realised very early, and a gas school for officers of all armies was organised at Leverkusen for training in protection. We cannot but regard it as significant that Leverkusen is also the site of the enormous Bayer[2] organic

[1] *My War Memories*, page 338.
[2] A branch of the great German dye combine, the Interessen Gemeinschaft, known as the I.G.

90

chemical works which played such a large part in poison gas production. The school dealt mainly with protection.

Early German Gas School.—Apparently, at the end of November, 1916, special gas staffs were created and attached provisionally to the head-quarters of formations entrusted with large scale gas operations. In addition, these staffs had the normal routine function of supervising inspection and instruction in gas warfare at the front. At about this time each regiment or larger unit was given a gas officer (gasschutzoffizier) with similar duties to those outlined above. In other words, the arrangement was generalised throughout the army. This officer was assisted by non-commissioned officers and men specially chosen for the purpose in the smaller units. The great need for these staffs is brought out in German official documents.

New Gas Regiments;—Gas Shell Experts.—In 1917 two new Pioneer battalions, the 37th and 38th respectively, were created for the express purpose of carrying out projector attacks. These developments in organisation, both advisory and combatant, led, at about this time, to the centralisation of the gas services at the front under a Kommandeur der Gas-truppen at G.H.Q. It would thus appear that the Germans achieved the centralisation of their gas services some months later than ourselves. Further developments in organisation, of which we are aware, were connected with two main tendencies in German gas warfare. In the first place, the vast employment of gas shell led the Germans to create special gas experts on the Divisional artillery staffs. We have this on the authority of an order by Ludendorff dated June 16th, 1918. This gas shell expert was not necessarily an imported specialist, but was usually a specially trained officer chosen from the staff in question. This was a very important move, for it gave the artillery a paternal interest in gas shell. This artillery specialist maintained a very close liaison with the Divisional Gas Officer.

Inspection of Protective Masks and Method.— The second tendency was towards stricter protective standards and inspection. The gas inspection centre at Berlin was given more responsibilities in the field and the protection of horses, dogs and carrier pigeons received great emphasis.

British Field Organisation;—"Breach" Organisations.—Our own field development followed very similar lines. The immediate need in April, 1915, was for organisations on the front to advise formations on temporary methods of protection, to ascertain quickly the nature of any new German chemical attack, and to provide special means of examining the treatment of the new kind of casualty. These were "breach organisations," so to speak, countering the immediate effects of enemy attacks while more comprehensive and permanent cadres were created to absorb them. The personnel of these breach organisations was largely composed of chemists already at the front who had in some cases taken part in the first German attacks. Efforts were soon on foot to mobilise British chemists for offensive purposes. So remote from the old army standards and training were the conceptions of the new scientific warfare, that there was no scientific cadre or outstanding scientific soldier to take over the direction and organisation of these matters at the front or at home. Accordingly, in June, 1915, Brigadier-General C. H. Foulkes, C.M.G., D.S.O. (then Major, R.E.) was given the difficult task of assembling and training an offensive gas formation, and acting as Gas Adviser to G.H.Q. The Special Companies thus created have already been referred to in quotations from despatches. In addition to this combatant personnel a number of specialists and advisory organisations came into being. Additional gas officers were appointed by various divisions, and chemical advisers by higher formations.

Central Laboratory.—A central laboratory was instituted at G.H.Q. under the late Colonel W. Watson, C.M.G., F.R.S., which did particularly valuable work in connection with the rapid identification of new enemy chemicals. With the development of gas shell, the chemical advisers included this subject in their province. Reference must also be made to the medical and physiological side.

New Type of Casualty.—After the introduction of gas warfare the army was always faced with the possibility that some entirely new chemical would produce a new type of casualty which would require special and sometimes unusual treatment. A new element was thus introduced into army medical work. The effects of a new gas used in large quantities on the front was often just as serious a threat to organisation as the sudden development of a strange epidemic. Reaction to meet these new conditions took the form of the development of medical research organisations at home, and of the appointment of a special medical and physiological advisory staff incorporated later in the Directorate of Gas Services. It was thus possible, after any enemy gas attack, and with little delay, to institute inquiries with regard to treatment of casualties, stimulate special investigations, and prepare for any reorganisation in personnel and equipment, and, in general, introduce satisfactory alert conditions throughout the medical organisation along the whole of the Allied front. In this connection the effective liaison between the medical specialists of the British and French armies must be mentioned.

Directorate of Gas Services.—These various services were centralised in the Directorate of Gas Services, in the Spring of 1916, under Major-General H. F. Thuillier, C.B., C.M.G., R.E. It is interesting to note that although in their rear organisations for research and supply the French preceded us in the adoption of a logical symmetrical arrangement, yet in the field we were the first to produce the centralised chemical warfare service which was so essential.

British Home Organisations;—The Royal Society.—After the battle of the Marne, Germany rapidly realised the need for scientific and industrial mobilisation for the new stage into which the war had passed. Many signs and definite statements by Falkenhayn and others in authority have shown us how this realisation found outlet in various schemes for research and production. The need for scientific attention to various war problems was also realised in England, and found expression in the mobilisation of prominent scientists by the Royal Society, which constituted a number of committees to deal with specific activities and to assist various Ministries or administrative government departments in connection with scientific matters.

Royal Society Chemical Sub-Committee.—The Chemical Sub-Committee included such prominent names as Lord Rayleigh, Sir William Ramsay and Sir Oliver Lodge. Retaliation, decided on early in May, 1915, was reflected in organisation. Lord Kitchener entrusted Colonel Jackson, C.M.G., R.E. (later Major-General Sir Louis Jackson, K.B.E., C.B., C.M.G., R.E.), then in charge of a fortification section of the War Office, with the task of examining and taking action on the possibilities of retaliation, and a liaison with the above chemical committee of the Royal Society was soon established. Protection became a part of the duties of the Medical Services and was placed under the direct control of Colonel, afterwards Sir William Horrocks, who became chairman of the specially appointed Anti-Gas Committee. Further, a little later, the Chemical Sub-Committee above referred to became an advisory body to Colonel Jackson. This was the origin of the Chemical Warfare Department, but it was destined to pass through many difficult and hampering transformations before reaching its final, more or less efficient and symmetrical form.

The Trench Warfare Department.—With the formation of the Ministry of Munitions late in May, 1915, Colonel Jackson's section was transferred to it. At this stage there was definite recognition of the absolute need of keeping chemical warfare research, design, and supply under one head. Probably this was the chief reason which prompted Lord Kitchener, then Secretary of State for War, to agree to the transference of this section to another Ministry, and consent to the birth of the Trench Warfare Department.

Scientific Advisory Committee;—Commercial Advisory Committee.—Even at this stage activities were growing and government organisation was found necessary to cover such functions as in Germany were rendered unnecessary by the existence of the I.G. It became clear that the new department would require strong permanent scientific advice, and this was found in the formation of the Scientific Advisory Committee. This included the most active members of the former relevant Royal Society Committee, amongst whom were Professor A. W. Crossley, the Secretary, and Professors H. B. Baker, J. F.

95

Thorpe, and Sir George Beilby, all of whom rendered great services in the later development of this new branch of warfare. A parallel Commercial Advisory Committee was appointed, composed of representatives of some of the leading manufacturers of the country.

Split Between Research and Supply.—We cannot follow in detail the many fluctuations experienced in the organisation of the department. They represent a constant struggle between a definitely expressed policy of centralisation and symmetry for supply and research, and circumstances imposed upon the department by the reorganisation and fusion of Ministries and departments. There were brief periods, notably at the commencement and in the final stages, when the desired centralised organisation was approached, but there were also periods when there was a complete split between research and supply with feeble and unsatisfactory liaison between the two. Speaking generally, the break between research and supply occurred in December, 1915, when the Trench Warfare Department was split up into two parts. These were the Trench Warfare Research Department, in which was included the Scientific Advisory Committee, and, shortly afterwards, changed its name to that of the Chemical Advisory Committee, and the Trench Warfare Supply Department. The relationships between those two departments remained practically unchanged until the formation of the Chemical Warfare Department in October, 1917. This statement must be qualified, however, by a reference to the services rendered by Professor, later Sir John Cadman, K.C.M.G., in bringing about this liaison, not only with supply in England, but also with that in France.

During the early period the Royal Society Committee of Physiology became active and was later very closely co-ordinated with the Chemical Warfare Department, as the Chemical Warfare Medical Committee.

Munitions Inventions . Department.—Another feature which is worthy of notice because it was common to Allied organisations other than the British, and because it formed part of the slow realisation of the essential unity of chemical warfare activities, was the duplication of effort by the Munitions Inventions Department. Suggestions which could only have value when considered as part of the definitely directed chemical warfare policy were constantly raised with the Inventions Department, but this difficulty was overcome later by the growing importance of chemical warfare and the effecting of a liaison between the two departments by Colonel Crossley.

Imperial College of Science.—During the early period the Imperial College of Science rendered great services by assisting in research. It continued to do so during the rest of the war, but was later associated with a large number of British university chemical and scientific departments in pursuing a huge programme of chemical warfare research. We can only make passing reference to the development of the training and experimental grounds which formed such an important part in assisting decisions on chemical warfare policy. The Porton ground, however, was a model of its kind, a pioneer amongst Allied experimental grounds, and a tribute to the creative and administrative efforts of Lt.-Colonel Crossley, C.M.G., C.B.E., who was its commandant from its inception to the end of the war.

The Chemical Warfare Department.—The growing importance of chemical warfare, the vigorous chemical initiative assumed by Germany in the summer of 1917, and various other reasons led to re-organisation of the Chemical Warfare services in this country in October, 1917, and the Chemical Warfare Department, under Major-General Thullier, formerly Director of Gas Services, B.E.F., was constituted. This reorganisation witnessed a great increase in research and other activities of the department and a still greater mobilisation of the chemists of the country. Although this change witnessed further centralisation by the incorporation of the Anti-Gas Department, thereby settling once and for all the inherent association between offensive and defensive research, a fact which had been apparent to many long before, yet it still ignored the fundamental connection between offensive research and supply. This had been recognised in French organisation as early as 1915, yet we did not reach the ideal solution even at the end of the war.

The Anti-Gas Department.—We have mentioned the origin of the Anti-Gas Department. Although separate in organisation from chemical warfare research, yet the remarkable work and personality of the late Lt.-Colonel E. F. Harrison, C.M.G., overcame the disadvantages by energetic liaison and a great capacity for the internal organisation. General Hartley has paid a tribute which we cannot refrain from repeating: "Colonel Harrison was one of the great discoveries of the war. It is often stated that he was the inventor of the box respirator, but this he would have been the first to deny. His great merit was as an organiser. He gathered round him an enthusiastic group of young chemists and physicists, and the box respirator represents the joint result of their researches, carried out under his inspiration and controlled by his admirable practicable judgment. He organised the manufacture of the respirator on a large scale, and it is a great testimony to his foresight and energy that in spite of all the difficulties of production, the supplies promised to France never failed. Fifty million respirators were produced by the department, and of these nineteen million were box respirators."

Anti-gas research was at first centred in the R.A.M. College, Millbank, and from the beginning of 1917 in the Physiological Institute, University College, London. The work done in research and production not only protected the whole of the British Army, but formed the backbone of American and a large part of Italian protection. Further, the sacrifices made in connection with this work are not sufficiently known. Numbers of young scientists sacrificed their health and sometimes life, in carrying out the critical tests upon which the safety of millions of Englishmen and Allies depended.

Designs Committee.—We cannot leave this branch of the subject without referring to the Chemical Warfare Designs Committee. An important trend in chemical warfare was its growing independence of the normal weapons of war, and its special requirements when adapted for use with both the normal and newer types. This tendency found expression in the above Committee under the direction of Professor Jocelyn Field Thorpe. The development of satisfactory chemical shell was an enormous problem, and the importance of entirely new forms of the chemical weapon will be brought out in dealing with the limitation of armaments.

French Organisation.—French development followed very similar lines.

From April 28th, 1915, a Commission of military representatives and scientists was organised under General Curmer. This gave place in June to a

Chemical Warfare Research Committee under M. Weiss, Directeur des Mines au Ministère des Traveaux Publics. In August, 1915, three special Committees were formed; one under M. Kling for problems from the front, whose organisation was responsible for a volume of exceedingly reliable identifications of enemy chemicals of great use to the Allies; another under M. Moureu for offensive research, whose brilliant organic investigation characterised later French developments, and the other under M. Vincent, for research on protection. But, in the meantime, the importance of gas shell was impressed upon the French and, on the 1st July, 1915, this organisation passed into M. Albert Thomas's new Ministère de L'Artillerie et des Munitions. Manufacture passed into the hands of the Directeur du Materiel Chimique de Guerre. In September, 1915, these sections were centralised under General Ozíl, attached to the same Ministry. General Ozil's service was strongly supported by a number of eminent French scientists, and achieved unusual success in the face of great practical difficulties.

A very close liaison was maintained with the army, and the initiative, energy, and devotion of all concerned cannot be too highly praised. In production alone the difficulties were enormous. There was no highly organised dye industry available. The pre-war German monopoly had seen to that. Elaborate organisations and continuous research work under difficult conditions were necessary to replace the smooth running normal activities of the great German dye combine. The salient points in French production are dealt with more fully in another chapter.

In research and protection French activities were no less handicapped and just as creditable. The protection of the French armies was largely achieved through the genius and tireless industry of Professor Paul Lebeau.

Quick to realise the need of retaliation against the new German weapon, the French developed their chemical offensive and defensive with characteristic élan and intuition. Contributing largely to Allied research, they took the lead in Inter-Allied co-operation and liaison, and their activities in this field were due to much worthier causes than mere geographical position.

Italian Development.—The Italians were alive to the importance of chemical warfare. World famous names such as those of Senator Paterno and Professor Villavecchia were associated with their organisation. Once again, however, although not lacking in invention and initiative, they were continually hampered by production, which imposed such grave disadvantages upon them as to endanger seriously the success of their campaign. The success of the great German offensive against Italy in the autumn of 1917 was largely ascribed to the German use of gas of such types and in such amounts that the Italian protective appliances were outmanœuvred. Further, in spite of the offensive qualities of the Italian gas organisation under Col. Penna, lack of supplies prevented large scale gas retaliation, so essential in maintaining gas morale.

Towards the end of the war, when the French and British production improved, and with the entry of America and the promise of supplies therefrom, it was possible to assist the Italians from Allied sources, and arrangements were made to supply them with the British Respirator, to assist them in the development of the Livens Projector, to supply large quantities of mustard and other gases, and to assist them in production. The use of the British box respirator was undoubtedly a great factor in repelling the Austrian offensive of June, 1918. Their experimental fields and research organisations were particularly well staffed, and, backed by production, Italian chemical genius would have been capable of producing very serious results.

Supply Organisations.—What a marked contrast between the organisation required for German and Allied chemical warfare production! Such organisation implies cadres and arrangements for co-operation with research organisations, for semi-scale work, commercial functions, priority, raw material supply, transport, and all their concomitants. In Germany, the self-contained dye industry simplified all these functions. The Government addressed itself to one producing organisation which was responsible for most of the relevant research. Whole Government departments were rendered unnecessary by this centralised production.

British Supply Organisation.—In England the situation was entirely different. Even before the advent of mustard gas the Government was compelled to apply to at least twenty contractors. The products required were foreign to the normal activities of many of these. They required assistance in raw materials, transport, technical methods, either the result of the work of other factories or of research. The latter again involved complex official organisation, cumbrous even if efficiently carried out. This at once introduced difficulties. The centre of gravity of supply was in government offices instead of in the centres of production. Much depended upon the co-ordination of the official departments. Quite apart from the Government plants finally engaged in chemical warfare production, more than fifty plants were used in private organisations, of which a very high percentage were entirely new.

Allied Handicaps.—The functions of the allied Government supply departments were or should have been much more than those of an individual negotiating a contract. Owing to the fact that these were new plants, and that the products were foreign to the production of many of the firms concerned, two alternatives had to be faced. Either the technical and service departments of each firm had to be considerably strengthened, or else a special organisation had to cover these functions by employing a considerable government technical and liaison personnel. For reasons of secrecy and general efficiency the latter procedure evolved, but neither represented the ideal solution.

The German Solution.—This was the German arrangement in which these functions were all embodied in the centralised producing organisation, the I.G. The German Government took the rôle of a pure contractor, the only additional function being the choice of product and method, a question of policy. This implied the existence of a Government experimental organisation, but purely for this purpose.

Departmental Difficulties.—The Allied task would have been much simpler if the only war weapon had been a chemical one, in which case an efficient organisation could have been decided upon at first, and need have suffered no very radical changes. As it was, however, the British supply organisation had to administer some seventy plants, which were really in private hands, and found its chief difficulties quite apart from the external perplexities of the problem. They arose in its relationships with other Government departments.

Allied Success Against Odds.—Taking a broad view of the case, although nobody who knew the facts could regard our poison gas production with anything but dismay, except in a few cases, yet the main feeling was one of amazement that we succeeded as well as we did with these entirely new substances. The whole story of chemical warfare supply amongst the Allies is one of devoted effort by all concerned, against overwhelming odds, and although the level of results was poor compared with Germany, yet we find here and there brilliant examples of Allied adaptability and tenacity amongst which the French development of mustard gas stands pre-eminent.

What we have already said about supply organisation may be summed up in one sentence. The Germans were already organised to produce. We had to create Government departments to administer a large number of plants in private hands, and they had to cope not only with the external difficulties of the situation but with the almost overwhelming difficulties of internal organisation. The chequered career of the British supply department provides a good example. The French and Americans suffered less than ourselves from these troubles, the latter having the benefit of the combined experience of the other Allies.

Allied Lack of Vision in Production.—A survey of the earliest supply organisation of this country reveals another difficulty which later events have obscured. Few people realised the developments which chemical warfare would produce. The early production of chemicals for gas warfare was grouped under some such designation as trench warfare stores, and graded in order of importance, from the point of view of supply organisation with catapults and spring guns, flame projectors and body shields! It is no unfair criticism to state that hard facts rather

104

than vision forced the importance of chemical warfare upon those responsible for munition production in the early stages of the war. Chemical warfare production remained under the Trench Warfare Supply Department for many months, where it was one of ten Trench Warfare sections. The vicissitudes of trench warfare supply are too numerous and complicated to be dealt with here, but chemical warfare supply has suffered accordingly.

British Lag in Organisation.—Examining Allied organisations, we find that the French and Americans approached this ideal solution more rapidly than ourselves, and we can trace in our own development a number of unsuccessful attempts to reach this centralised control, although the last configuration, under the direction of Major-General H. F. Thuillier, was the nearest approach. French organisation for supply provides another example of their national characteristic of logical thinking and love of symmetry. As early as September, 1915, the French centralised their research organisation, the Inspection des Études et Expérience Chimiques, and their supply organisation, the Direction du Matériel Chimique de Guerre, in their Service Chimique de Guerre under General Ozil.

French and American Characteristics.—Their early concentration on gas shell shows that this symmetrical organisation was due not only to the above characteristic but also to vision in war development. American supply organisation again provides evidence of the national characteristic. They had no I.G. but they had plenty of money and material, and the total of Allied experience in production. They therefore proceeded at once to build an enormous producing centre known as Edgewood Arsenal. We refer to this later. The tremendous potentialities of this Arsenal will readily be seen, although they did not become effective during the war.

It would be poor testimony to the tremendous efforts and sacrifices made by the various firms and officials connected with chemical warfare to leave the matter at this stage, or to make a minute analysis of the different internal causes for lack of success. We may say that although the efforts of all concerned were beyond praise, yet they were so initially handicapped that it was practically impossible even to approach the German efficiency. In France and England we were suffering from the faults of past years, our lack of attention to the application of science to industry. The Americans would also have suffered, for they were in the same plight, but they adopted the drastic solution of Edgewood Arsenal. As we show later, however, this solution was really only a very necessary and valuable attempt to treat the symptom rather than the disease. We cannot regard the problem as settled for any of these countries. If it is, then the outlook is very poor.

Inter-Allied Chemical Warfare Liaison.—Chemical warfare offered, in theory, a splendid opportunity for co-ordination amongst the Allies. The new methods, unhampered by tradition, seemed, at first sight, admirably suited for exploitation against the enemy by an allied Generalissimo and staff. Co-ordination never reached this stage, although strong liaison organisations were developed. Inter-allied research conferences occurred periodically in Paris, where decisions for co-operation were taken after full discussion of allied work. The continuity of these relationships was maintained by an active secretariat on which each ally was represented. The contact, so close between actual allied scientific workers in this field, became less evident in the application of their results to field warfare, for several reasons. In the first place, close scientific contact in research was replaced by the actual field realtionships of the armies, and, as is well known, the central inter-allied command did not materialise until the spring of 1918, and even then it was only possible to apply the new principle to the actual battlefield. The traditional differences between the methods of the different services of each ally still existed to a large extent, and they found expression in type of armament, equipment, and military standards, such as, for example, gun calibres and shell design, to which chemical warfare had to conform. No inter-allied gas mask materialised, although this would have been of inestimable advantage. Probably the example of most complete co-ordination occurred on the supply side, where absence of the above traditional difficulties and the crying need to make the most of available raw materials compelled a very close co-ordination.

Inter-Allied Supply.—The writer was responsible for initiating, in 1917, an Inter-Allied Chemical Supply Committee, whose function was to pool effectively the allied raw materials, and to arrange their distribution in accordance with allied programmes, the exchange of which implied a considerable step. Later this Committee became one of a number, similarly constituted, forming part of the Inter-Allied Munitions Council.

Thinking over the difficulties of the inter-allied supply, now that the emergenices of the situation have passed, an important contrast emerges. After three years of war, and although protected by the powerful arm of the blockade, we were still resorting, for chemical warfare supply, to measures which, compared with the German methods, were complicated, clumsy, and inefficient. This was, in a sense, forced upon us by the number of the allies, and the fact that they held the outer lines. But it is easily forgotten that Germany also had a number of allies, and that German supply organisation was sufficient to feed them all.

Nature of Chemical Warfare Research.—So much has been vaguely said, and is vaguely known, about research in chemical warfare that a brief analysis will be of value.

Discovery of New Substances.—Research for this purpose has a number of very distinct functions. The most obvious is the discovery of new substances. But there are others in connection with which research work represents a much greater volume. Very few new substances which found valuable application during the war were revealed by chemical warfare research. The bulk of the important substances were already known as such, although their importance for war was probably not realised. It is most important to emphasise the fact that even in the future, should there be no direct attempts to reveal new chemical warfare substances, they will undoubtedly arise as a normal outcome of research, even if, without exception, every chemist in the world became a most pronounced pacifist. A valuable substance once discovered or decided upon, however, whole series of research investigations become necessary.

Technical Method of Preparation; — Filling Problem;—Protection;—Half-Scale Investigation. —The substance must be prepared in the most efficient manner for manufacture, which may not be the mode of its discovery. It must be used in shells, cylinders, or some other war chemical device. Each device represents a different filling problem, different difficulties with regard to contact of the war chemical and the envelope of the container. If a projectile is in question the ballistics become of importance. More important than any of these, except production, is the question of protection. It is axiomatic that an army proposing to use a new offensive chemical must be protected against it. It may, therefore, be necessary to modify the existing mask or protective appliance, or to create an entirely new one. If research reveals the necessity for the latter course of action it may provide sufficient reason for abandoning the substance. In addition, according to productive difficulties, it may be necessary to undertake comprehensive and very expensive research on half-scale methods for production. It is impossible in many cases to proceed directly from the laboratory process to large scale manufacture without serious risk of failure.

Two Classes of Research.—Broadly, these research functions form two classes, those concerned with policy and approval of a substance and those concerned with work which follows automatically upon such approval. There must be, of course, a certain amount of overlapping and liaison between the two classes.

Herein lay one of the great advantages enjoyed by the Germans. Their great producing organisation, the I.G., was able to take over automatically certain of these research functions, in particular all those with regard to preparation and production, even of protective appliances. The Government reserved what we have called the policy functions, and was responsible, we assume, for the mass of physiological and design research which must always precede approval or a decision on policy.

Signs were not lacking, further, that the I.G. was even employed on certain occasions for this latter type of research.

Conclusion.—From the facts at our disposal there can be no doubt that the total material facilities at the disposal of the Allies for chemical warfare investigation were considerably more extensive and expensive than those of Germany with the one notable exception of trained technical organic chemists. It is very doubtful whether the German field experiments were as largely provided for as those of the Allies. When we think of the French grounds at Versailles and Entressin, the British at Porton, the American grounds in France and in America, and the Italian organisations, there can hardly be any doubt that the total German facilities were much smaller. Under the actual circumstances of the war, however, it was difficult to develop more co-operation than was possible by a very close liaison. The fact that all the experimental developments from these grounds required special modification to meet the peculiar needs of artillery and other equipment for each ally, prevented the adoption of uniform types of projectile or other appliances. Even uniform shell marking was found impracticable.

The "Outer and Inner Lines."—The Allied situation compelled the multiplication of cumbersome organisations in the different countries. Lack of a strong organic chemical industry placed each ally at a considerable disadvantage, compared with Germany, in the development of such organisations. Using a strategic comparison, we can say that Germany not only possessed the "inner lines" in the chemical war, but an exceptionally efficient system to exploit them, in the shape of the great I.G.

CHAPTER VI

THE STRUGGLE FOR THE INITIATIVE

Meaning of the Chemical Initiative.—The German invasion of Belgium in 1914 was a direct appeal to the critical factor of surprise in war. By disregarding their pledge, a "scrap of paper," they automatically introduced into this attack the elements of military surprise. We, the enemy, were unprepared, and a complete rearrangement of dispositions became necessary.

A recent writer has admirably summarised the facts.[1]

"Germany began the war on the Western front before it was declared, and on 1-2 August, German cavalry crossed the French frontier between Luxembourg and Switzerland at three points in the direction of Longwy, Luneville, and Belfort. But these were only feints designed to prolong the delusion that Germany would attack on the only front legitimately open to warfare and to delay the reconstruction of the French defence required to meet the real offensive. The reasons for German strategy were conclusive to the General Staff, and they were frankly explained by Bethmann-Hollweg to the British Ambassador. There was no time to lose if France was to be defeated before an effective Russian move, and time would be lost by a frontal

[1] A. F. Pollard. *A Short History of the Great War.* Methuen, 1920.

attack. The best railways and roads from Berlin to Paris ran through Belgium; the Vosges protected more than half of the French frontier south of Luxembourg, Belfort defended the narrow gap between them and Switzerland, and even the wider thirty miles' gap between the northern slopes of the Vosges and Luxembourg was too narrow for the deployment of Germany's strength; the way was also barred by the elaborate fortifications of Verdun, Toul, and Nancy. Strategy pointed conclusively to the Belgium route, and its advantages were clinched by the fact that France was relying on the illusory scrap of paper."

The first German cloud gas attack was the second attempt to gain the decisive initiative, by the unauthorised use of a surprise of an entirely different nature.

Modern writers are at great pains to establish how the world war, although leaving the final function of the infantry unchanged, rendered them and their staff subservient to mass munition production. Mr. H. G. Wells explains this to the Kaiser in a delightful imaginary interview between that august person and an hypothetical manufacturer.[1] Professor Pollard tells us how, when the first German surprise had failed, the war became "a test of endurance rather than generalship." We will leave a clear field for any military challenge to such a point of view. Our objection is that it is not fully developed. The war was still a test of generalship, that of directed production. This war has shown, and future wars may unfortunately confirm, that the type and secrecy of production is as important as its volume. There will still be the purely military surprise and manœuvre, but

[1] *War and the Future.* Cassel, 1917.

superimposed, co-ordinated, and sometimes predominant will appear the technical surprise, the result of the generalship of production.

Such a surprise is achieved by the sudden introduction on a large scale of some entirely new war weapon, capable of achieving a strategic or tactical objective in an unsuspected manner.

Although the general idea of this second type of surprise existed before the war, particularly in naval warfare, it required the coincidence of the Great European War and modern scientific development to demonstrate its great importance on land.

Thus the first German gas attack found the opposing troops entirely unprotected, not merely through the absence of a mask, but in training and technical discipline. The case is quoted of an indignant gassed soldier who, in an early gas attack, when reproached for not protecting himself, thereupon opened his tunic and revealed a mask firmly tied round his chest! It is a far cry from such a case to the inculcation of strict gas discipline into an army of millions. The attack reaped the corresponding results in casualties and morale. It found the opposing medical services unequipped, not only to treat the new type of casualty, but even to determine its nature rapidly and efficiently. In short, it found the enemy utterly unprepared, either in theory or practice, to counter its effect. The importance of this second type of surprise lies in its peculiar potentialities. It may affect a given military result with an extraordinarily small expenditure of material, energy, and eventually human life, when compared with the older military weapons. Chemical warfare is a weapon, par excellence, to achieve this second type of surprise. Therein lies its chief importance.

As a result, the history of chemical warfare becomes one of continual attempts, on both sides, to achieve surprise and to counter it by some accurate forecast in protective methods. It is a struggle for the initiative.

More than this, as the use of chemical warfare becomes an organic part of operations, as it did during the war, these operations become correspondingly dependent upon conditions imposed by the chemical campaign. One can imagine the case of an army unprotected against a new gas, aware that the enemy is ready to employ the latter, compelled to postpone some huge offensive until its protective methods were equal to countering the new chemical. General Fries, the American authority, states, in reference to mustard gas, and the Northern offensives in 1917: "It is no disparagement of the British, nor of any one else, to say that they held up the date of their attack for two weeks pending further investigations into the effects of this new gas." Ludendorff, referring to the German offensive in March, 1918, tells us, "Our artillery relied on gas for its effect. Up till the morning of the 20th strength and direction (of the wind) were by no means favourable, and it seemed almost necessary to put off the attack." Such a point becomes of greater importance as the influence of other arms decreases. If we assume international arrangements for the limitation of other types of armament in the future, chemical warfare at once stands out as decisive.

Controlling Factors;—Rapid Manufacture.—Certain well-defined factors hold a controlling position in the chemical initiative. Before any chemical discovery can be used for surprise on the front a second step must occur; this is large scale manufacture. This period is vital to surprise. Success in chemical warfare is largely dependent on secrecy, which means achieving production in the shortest possible time, and this is particularly important at the commencement of hostilities. Throughout the war the Germans possessed this advantage and, in the future, unless certain steps are taken, it will be theirs again. A very simple example will suffice to show the importance of the combination of these two factors. Let us assume the not remote possibility that Germany had refrained from using poison gas until she had reached the stage of development which existed at the time of her 1918 offensive. There is little room for doubt that the big scale use of cloud attacks which would then have been available, and of shell gas, in particular mustard gas, would have achieved decisive success. The Allies would have been totally unprotected, the moral effect would have been enormous, and, even if we ignore the latter, the number of casualties would have produced a gap the size of which was only dependent on German wishes.

Rapid Identification Essential.—It is important to remember, however, that once a chemical campaign has commenced, certain factors may militate against any lengthy retention of the initiative by either party. Organisations develop whose function is to ascertain the nature of new enemy chemical devices so that protective research and production can commence with the minimum delay. This assumes the existence of a protective appliance and organisation. The very efficient collaboration of the British Central Laboratory in France for the examination of new gas shell with the French organisation centred in Paris provides numerous examples of the functioning of this safeguard. No time was lost in identifying the nature of the various chemicals employed by Germany in her shell fillings. Speed was vital. The use of a new type of chemical in shell, bomb, or other contrivance, in any sector of the front, on whatever scale, however small, was reported without delay. Then followed instantaneous collection and examination, after which all front line formations, other formations, allies, and rear organisations were expeditiously warned. The harmless trial flight of the few shell of a new type might be followed by the use of hundreds of thousands in a deadly attack one hundred miles away or on another allied front. Not only were captured offensive contrivances of value for this purpose, but the rapid examination of new enemy masks was of prime importance, for it could be assumed that the enemy would be protected against his own surprises in store for others.

Attempts to ascertain the enemy's gas activities were not confined to examining captured material after their first use. Raids and artillery fire were both used to obtain intelligence regarding preparations, or to break up the gas emplacements. The Germans have provided us with a particularly gallant and interesting attempt.

Near Nieuport the front penetrated a region inundated by the Belgians during the desperate German offensives of 1914. The trench system, winding through a mile or so of sand dunes, passed in a southeasterly direction through the marshy sector known as Lombartzyde. Here the bogged front lines were intersected by the Yser canal, the German front trench being some 80 yards away. Allied gas was installed in the Lombartzyde and neighbouring sectors ready for discharge on the first favourable opportunity. For some reason or other the Germans suspected this, and at night a raiding party swam down the ice cold Yser, and, negotiating the submerged wire, landed in the Allied support line. Stunning the sentry with a bomb which, fortunately, refused to explode, they proceeded to the front line to seek gas emplacements. Either through unexpected disturbance, or for some other reason, they were compelled to leave before completing their inspection, and successfully swam the Yser canal back to their own trenches. This hazardous enterprise represents but one of many raids whose function it was to ascertain the presence of enemy gas.

Propaganda and Morale.—Another factor intended to facilitate the attainment of the chemical initiative was the German use of propaganda. Rumours, reflected in the Press, were often current at the Front, at home, and in neutral countries, that some particularly fiendish chemical contrivance was about to be launched against the Allies by Germany. Thus, in January and February of 1916, vigorous propaganda activity of this kind in Switzerland preceded the great German offensive at Verdun. The new gas was heralded by fantastic stories. Certain death was threatened for all within one hundred yards of the shell burst. The origin of the report was traced to various sources. In one case rumours concerned a conscientious worker in a German factory, desirous of warning the French through Swiss friends, in other cases German scientists were reported to be influencing Francophile neutrals in order that they might warn the French. But an analysis of the propaganda reveals something more than its sensational nature. The information arrived at well-defined periods, which usually preceded the actual use of a new gas or chemical device by Germany. But when the actual effort is compared with the prophecy we find that in no case was there any real clue as to the nature of the gas. Thus, before the use of phosgene by the Germans at the end of 1915, definite reports reached the Allies regarding the projected use of at least ten new gases by Germany, which were described not only chemically, but as being colourless, odourless, powerful, blinding, and instantaneously deadly! No such volume of propaganda was experienced before the first German cloud attack at Ypres. Indeed, one would not have expected it, for the mere fact of the use of cloud gas was then new to war, and of military value.

This propaganda was not without its effect, and, but for the excellent Allied gas discipline, would have been an effective precursor to the gas itself. Cases were not absent, at the Battle of Loos, for example, in which the German use of lachrymators found British soldiers so mentally unprepared, or rather let us say " prepared " by propaganda, as to spread ridiculous rumours on the battlefield as to the all-powerful nature of the new German gas shell. These were, in fact, bursting a few yards away, with no more serious results than lachrymation and vomiting. The extended use of shell gas by the Germans in the summer of 1916 was again preceded by intensive propaganda during the early months of that year, in which the promise of prussic acid was prominent. The influence of a name is very curious. Prussic acid probably accounted for fewer casualties than any other gas. This fact became apparent with the increasing use of the French Vincennite, which contained prussic acid. Yet German propaganda redoubled its efforts as time went on to inspire fear in the Allied soldiers by the threat to use prussic acid. It is clear that armies cannot abandon gas discipline, and that an important factor in strengthening this discipline is a wise distribution of gas knowledge. The use of mustard gas and newer shell gases in 1917 was again preceded by a burst of propaganda. In this period we find the first reference to long-range gas shell and aircraft gas bomb, and, curiously enough, a certain amount of propaganda with regard to a blinding chemical, which partially described mustard gas.

As further confirmation of the General Staff origin of this propaganda we find that the 1918 outburst occurred two or three months earlier in the year than in 1917. This was accounted for, no doubt, by its intended influence upon Allied morale in the great German offensive of early 1918. This last wave of propaganda includes one very interesting example. It is better known than other cases through its association with the International Red Cross at Geneva. This body represented in February, 1918, that Germany was about to use a really terrible gas which would have such disastrous effects that it was absolutely essential to make a last attempt to get both sides to abandon gas warfare. The official wire reads as follows:—" Protest of International Red Cross against the use of Poison Gas. I have received private letter from Monsieur X., President of International Red Cross, which I think that I ought to lay before you. He says that Red Cross were induced to make protest by what they had heard of new gas Germans are preparing although Red Cross understands that the Allies are aware of the gas and are taking their precautions. As they did not wish to draw an indictment of Germany they appealed to both groups of belligerents to pledge themselves not to use this weapon. Red Cross asks whether the Entente leaders through Inter-Allied Council at Versailles could not make a loud declaration which would reach the peoples of the Central Empires as well as their rulers, pledging themselves not to use such gas on condition that the two Emperors similarly bind themselves not to employ it. If the latter refuse, all the guilt will rest with them." Although there can be no doubt that the International Red Cross and the Swiss involved in this move were absolutely bona fide, yet whoever was responsible for initiating the move on the German side played his hand very well. If, as actually occurred, the protest did not result in the cessation of gas hostilities, it still served its purpose as propaganda aimed at Allied morale. Knowing his dispositions for gas defence, and our own offensive preparations, it is probable that the enemy was willing to withdraw before being over-whelmed by Allied and American production. After three years of costly improvised production by the Allies, Germany could no longer securely enjoy the fruits of the initiative provided by the plants and factories of the I.G.

Peculiar Peace-Time Danger.—There can be no doubt therefore that the mere contact of two armies during war acts as a check against the decisive use of chemical warfare, except in the very early stages. During peace this contact will be practically non-existent, and it would be possible for any country so to diverge in its lines of research and discovery that, given rapid means of production, it could repeat the German surprise of 1915, this time with decisive results. Should such a nation possess a monopoly in the means of rapid production, the world is practically at her mercy. Should she be prepared to break her word, the usual means of controlling disarmament are impotent against these developments.

War Fluctuations of Initiative.—In the light of the above remarks the fluctuations in the initiative during the recent war are very significant. The first marked feature was the development of British and Allied protection to counter the enemy attacks which would presumably follow the first German use of cloud gas. Immediately after the German chemical surprise, and while the Allies were still undecided whether to retaliate, work proceeded feverishly on the development of some form of protection for the hitherto unprotected soldier. In response to Lord Kitchener's dramatic appeal to the women of England and France, masks were sent to France in sufficient quantity within a few days. They were of a very primitive type, and consisted of a pad of cotton wool impregnated with certain chemicals, to be held in place over the mouth, which was superseded, in May, by a very similar contrivance, slightly more efficient with regard to the length of time of protection. Dr. Haldane and certain other prominent chemists and physiologists worked on the different improvised types. With this feeble protection, or, in the first case, with none at all, our armies had to face the first German cloud gas attacks.

The idea of the gas helmet which covered the whole head was brought to England by Captain Macpherson of the Newfoundland Corps, early in May. Suitably impregnated, it made satisfactory tests. The helmet type of respirator made of flannel was first tested in the Anti-Gas laboratories on May 10th, 1915, and was a great success compared with previously suggested types. Arrangements for its manufacture were accordingly made, and this began in June, 1915. This protective device consisted of a flannel helmet with a celluloid film eyepiece, and was called the hypo helmet. The fabric was impregnated with the same solution as the cotton waste pads described above, the dipping being carried out largely at Oxford Works, but partly in the Royal Army Clothing Department, Pimlico. Its manufacture was continued until September, 1915, about two and a half millions being made in all. From June, 1915, we never really lost the initiative in the matter of defence, although, at different times, the struggle was very intense. It was this helmet, with the modified phenate impregnation, which, known as the P. helmet, formed the first line of defence against the probable employment of phosgene by Germany. It became known as the "Tube Helmet" when fitted with a mouthpiece for exhaled air, and, in this form, countered the formidable enemy phosgene attack in December, 1915. The later addition of hexamine, suggested from Russia, greatly improved the efficiency against phosgene and led to the P.H. helmet, which was issued from January, 1916. It was not withdrawn until February, 1918, but in the later stages was used as a second line of defence. The magnitude of this manœuvring for protection can be judged from the facts that two and a half millions of hypo helmets, nine millions of P. helmets, and fourteen millions of P.H. helmets were issued during the campaign.

There is no doubt that this early period, however, was a very costly experiment on the use of the different masks, the success of which involved the loss of numbers of men who were compelled, through reasons of supply or uncertain design, to use the less efficient types. In one case, for example, the trial of mica eyepieces rendered otherwise efficient masks absolutely useless by breaking, and caused losses. We cannot afford to repeat such experiments in future. Failure to develop protective appliances fatally implies large-scale experiments in future wars in which unnecessary loss of life is bound to occur. If steady research in peace can diminish this possible loss, shall it be stopped?

The urgency of these developments can be understood from a case quoted by General Hartley:[1] "A certain modification of the respirator was considered necessary in France, and officers were sent home to explain what was needed. Within forty-eight hours of their arrival arrangements were made to modify the respirators, and within a few weeks the fighting troops had been re-equipped with the new pattern. Less than three months after the change had been recommended three attacks were made by the Germans which would certainly have had very serious consequences if our troops had not been in possession of the improved respirator, as the older pattern would not have withstood the concentration of gas employed. This was only one of many changes that were made in the respirator to meet new developments."

How urgent was the need for these developments? It was vital. Here is a case showing frightful losses sustained by partially or inefficiently protected troops. Between May and July of 1915 the Germans made at least three cloud gas attacks upon the Russians, immediately west of Warsaw. In all these attacks, taken together, gas was discharged for a total time of not more than one hour, and they were all practically from the same position, on a front of about six miles. The affair seems relatively small, yet what was the result? The Russians lost

[1] Report before the British Association, 1919.

not less than 5000 dead on the field, and their total casualties were of the order of 25,000 officers and men. A Siberian regiment had, before the last attack, a ration strength of about forty officers and 4000 men. This was reduced by a twenty minutes gas discharge to four officers and four hundred men. No other weapon could have reproduced, under the most favourable conditions for its use, in as many days, what gas was able to do in as many minutes.

Although our protection had countered the later German attacks with cloud gas, yet it threatened to fail to meet the situation created by the use of a variety of organic chemicals in shell. In order to counter the use of lachrymatory compounds by the enemy, compounds which penetrated the helmet insufficiently to cause serious casualties but sufficiently to hamper the individual by lachrymation, goggles were introduced in which the eyes were protected by rims of rubber sponge. This remedied the weakness of the P.H. helmet and produced the P.H.G. helmet, of which more than one and a half millions were issued during 1916-1917.

Towards the end of 1915 the standard protection was the P. and P.H. helmet, but the use of lachrymators compelled us to use the P.H.G. Even this helmet was not satisfactory against the high concentrations of phosgene or lachrymators, and after much research the opinion gained ground that further development must be on other lines. In addition, the need for a more general form of protection was emphasised by the German adoption of a mask of cartridge design. In other words, the fabric of the helmet, or facial portion of the mask, was made impermeable, and the filtration of the poisoned air occurred through a cartridge, or filtering box, attached to the fabric in the form of a snout. The cartridge provided a much greater protective range and capacity. It was clear that such German protection was evidence of their plans for the further use of gas. The new German cartridge mask appeared in the autumn of 1915. Doctor H. Pick, reviewing German protective measures in Schwarte's book, enumerates the various desiderata of the ideal mask and explains, "It was only our early recognition of these requirements which gave us an advantage over the enemy from the first in the sphere of defensive measures against gas, and which spared us from having to undertake radical alterations in the apparatus as the English, French, and Russians had to do more than once." This early adoption of a comprehensive view on protection by Germany is a testimony to both German thoroughness and their definite intention to proceed with a vigorous chemical war. The latter is not mere inference, for it is borne out by the dates at which they commenced production in their dye factories. Further, even if the German cartridge mask was only decided upon after Loos, which is not probable, our feeble reply in that battle would hardly have justified such a radical advance in protection.

It was thus forecasted that not only would new ranges of compounds be employed which it would be most difficult to counter individually, but aggressive methods would arise, either entirely new or modifications of the cloud method, which would enable much higher concentrations to be obtained than those in evidence hitherto. Accordingly the first type of the well-known British Box Respirator was designed, giving a big capacity of highly efficient filtering material, or granule, contained in a canister, with an improved face-piece and breathing arrangements. Without going into details, it may be said

125

that Colonel Harrison and Major Lambert were associated with a number of other enthusiastic workers in developing the Box Respirator.

Here again the question of chemical supply threatened to influence our retention of the initiative. Without going into the development of the granule in the respirator, the supply of potassium permanganate was of prime importance, and the country was woefully deficient in the production of this substance. The determined efforts of British manufacturers overcame this difficulty. It was now possible to work on general lines for the improvement of this canister to increase its protective range, and to modify the canister specifically in accordance with intelligence as to what the enemy had recently done or was about to do. In this way, and successively, the army was successfully protected against the higher concentrations employed and the newer substances introduced. The issue of the large Box Respirator commenced in February, 1916. It was replaced by the small Box Respirator which came out in August, 1916, and of which over sixteen millions had been issued before the signing of the Armistice. At one time over a quarter of a million small Box Respirators were produced weekly. The chief modifications were the use of a smaller box or canister, the margin of protection being unnecessarily large in the former type.

It became necessary in the spring of 1917 to provide more efficient protection against irritating smokes which tended to penetrate the respirator as minute particles, and the first form consisted in the use of two layers of cotton wadding in the canister of the small Box Respirator. The use of Blue Cross compounds by Germany in the summer of 1917 rendered this matter more urgent, and a special filter jacket was designed which fitted round the Small Box Respirator. A million were made and sent to France. Developments proceeded on these lines. Altogether, more than fifty million masks and respirators of different kinds were manufactured by the British Anti-Gas Department for our own and Allied armies.

We thus have some idea of the importance of protection in chemical warfare and of the absolutely imperative need of deciding whether or no work on protection must go on. There can be no doubt as to the answer to this question. It is not only in the interest of the army, whether a League of Nations or a national army, but also in those of the civil population.

The Tense Protective Struggle.—Few people realise how the development of Allied and enemy gas masks and protective measures was forced upon each side in a number of critical steps. At each of these, had research and production been unequal to the task, the armies would have found themselves more uncovered and exposed than if the whole trench and dug-out system had been suddenly rendered unusable in some peculiar way, thus removing cover from high explosive and shrapnel, rifle, and machine-gun fire. The army has an apt expression. An officer or man parading incompletely equipped is dubbed "half naked." To be within reach of enemy gas without a mask was true nakedness. A modern army without a gas mask is much more helpless and beaten than one without boots. More than this, it must be clearly understood that a gas mask of efficient design and production will remain of very little use unless, supported by comprehensive research which, itself, gains enormously in efficiency if related to enemy offensive activities.

The Riddle of the Rhine

The German Mask.—Consider the German mask for a moment. We have seen how Germany adopted the canister drum or cartridge form before any of the other belligerents, and in good time to protect her own men against their own use of phosgene, at the end of 1915. Indeed, Germany probably held up the use of phosgene until her own protection against it was developed, although Schwarte's book claims that the German mask issue in 1915 was mainly a protection against chlorine. The filling consisted of some such material as powdered pumice-stone saturated with a solution of potash, and powdered over with fine absorbent charcoal in order to protect against organic irritants and phosgene. These were the familiar one-layer drums. Then came the British concentrated cloud gas offensive in the summer of 1916, which undoubtedly found the German mask unequal to some of the higher concentrations which were obtained under most favourable conditions. The Gas Officer of the Sixth German Army stated in a document issued in November, 1916: "Considerable losses were caused by the gas attacks which have taken place latterly. The casualties were mainly due to the men being surprised in dugouts, to the neglect of gas discipline, masks not being at hand, to faulty masks, and to the use of old pattern drums *which could not afford protection against the type of gas employed by the enemy.*"— (The italics are our own.—V.L.)

Evidence is found in the introduction of the German three-layer drum in the autumn of 1916. An army does not undertake the manufacture of millions of new appliances without very good reason. This new drum was specially aimed at phosgene protection. The middle layer consisted of granulated absorbent charcoal, which had the

128

property of absorbing large quantities of organic irritants and phosgene. In the three-layer drum the latter gas was adequately guarded against for most field purposes, although we have reason to believe that the German staff was always apprehensive, and German soldiers suspicious of the actual penetration of their mask obtained in the immediate locality of projector discharges.

Dr. Pick explains in Schwarte's book what is already well known, that the charcoal layer has a wide, "non-specific effect, and it retains almost all materials of which the molecular weight is not too small, even if very strongly neutral in character (as, for example, chlorpicrin)." He goes on to say "the progressive development of gas warfare led to the use of these very materials, whilst substances with acid properties, such as chlorine, fell more and more into disuse. The three-layer drum went through all sorts of changes in consequence. When the use of chlorpicrin mixtures gained in importance in 1917, the layer of charcoal was increased at the expense of the other two layers. This stage of development ended in 1918, when the other layers were done away with altogether, and the entire three sets were filled with 'A' charcoal." " 'A' charcoal was a particularly efficient form. We learn from the same source that the increased protection against phosgene was very welcome to the Germans in view of the danger arising from gas projector attacks. Further, the capacity for absorption of the German charcoal was never equalled by any of foreign production." This was certainly true for the greater part of the war. But Dr. Pick continues, in a sentence which is full of significance: "In consequence of the high quality of the drum's absorption, we were able to carry on to the end of the war with a drum of

129

relatively small proportions." This point is so important as to demand further explanation.

Enforced German Modifications.—The most important disadvantage of a gas mask is its resistance to breathing. Men undertaking arduous and dangerous duties in the presence of gas must wear a mask, but they cannot undertake these duties if their breathing is seriously interfered with. This is particularly so in trench engineering and in the heavy work of the artillery. Now the resistance depends, for a given type of filling, upon the area of the cross-section of the drum. Breathing will be easier through a very large area than through a very small one. The British appliance was a frank admission that, with its filling, a large drum was necessary, so large that the weight of it could not be borne by the mask itself, but by attachment to the chest, the actual mask being connected with the drum or box by a flexible rubber tube. But the Germans adopted from the beginning a form of protective appliance in which the drum or cartridge was attached to and supported by the mask. In other words, their development was limited by the weight of their drum, unless they completely changed their type on British lines. It is quite clear that they realised this, for Doctor Pick tells us, referring to the large size of the British box: "For this reason the weight of the box is so great that it is no longer possible to attach it directly to the mask. It is, therefore, carried on the chest and joined to the mouthpiece of the mask by a flexible tube."

The development of British cloud gas compelled the Germans so to modify their filling that the resistance to breathing increased considerably. They countered this, however, by introducing an exceedingly active charcoal, realising that the weight of their drum had already reached the limit possible with that type of apparatus, and that they could not, therefore, get better breathing capacity by increasing its size. When, however, the Blue Cross compounds were introduced, it was necessary for both armies to take special precautions. These precautions involved introducing a layer of filtering material into the canister or drum. Dr. Pick tells us: "When the material of the Blue Cross type became of greater importance, a supplementary apparatus had to be issued. A thin disc filter prepared by a special method from threads of cotton was fastened to the tube of the drum by means of a spring lid. This arrangement provided adequate protection against materials of the Blue Cross type used by the enemy, as, for instance, stannic chloride, whilst the German Blue Cross gas, which was more penetrating, was only retained to a moderate degree." This is a direct admission that, in order to counter the Allied use of Blue Cross gas, further filtering arrangements would have been necessary. But the resistance to breathing of the German apparatus was already strained to the utmost. It is exceedingly improbable that the Germans, having already reached the limit of size of the canister or drum, and being unable to obtain better breathing by increase in size, could have introduced any such device without carrying their resistance beyond the possible limit. In other words, the use of Blue Cross by the Allies would have compelled them to adopt the British type of apparatus, that is, a bigger box supported by the chest and connected to the mask by a flexible rubber tube. This would have led them into an *impasse.*

Shortage of Rubber.—We know how, in the beginning of 1917, they were compelled to substitute leather in the substance of the mask. Dr. Pick

131

admits that this was due to lack of raw material, rubber, and there are many other signs that this was so. Although leather was not altogether a bad substitute for this purpose, rubber would have been essential for the flexible tube, and the millions required to refit the army would have completely broken the German rubber resources. Many facts, including their feverish development of synthetic rubber, small quantities of which they obtained at enormous cost, go to prove this conclusion. The submarine, *Deutschland,* returning to Germany in 1916, from its historic trip to America, carried shipments of the most sorely needed commodities, including large quantities of raw rubber. Stringent measures were adopted later to collect waste rubber and prevent its use for such purposes as billiard tables and tyres for private vehicles. The first naval expedition to Baltic ports after the Armistice found the hospitals in a pitiable plight for lack of rubber. The Germans were being driven into an impossible position. In other words, the Allies, by a proper use of Blue Cross compounds, could have regained the gas initiative. There is no doubt that they were within a few months of doing so. Once again we see the importance of production. Lack of raw materials for protective purposes was endangering the German position, but delay in offensive production by the Allies removed that danger. Although their pressing need was obvious, the Blue Cross arsenic compounds were not available. The chemical war involves manœuvring for position just as definitely as the older forms, but in it production, formerly a routine activity, assumes critical strategic importance.

Gas Discipline.—This constant vigilance against enemy surprise imposed more conditions upon the troops than the permanent adoption of a protective

THE LIVENS PROJECTOR—III.
Explosion of Livens bombs on the objective.

To face page 133

appliance which, in itself, was a very big thing. Given the mask, the army had to be taught how and when to use it. A gas sense had to be developed which ensured rapid use of the mask at the right time with the least hampering of operations. Gas discipline thus became one of the most important features of general training, a feature which can never be abandoned by the armies of civilised nations in the future without disastrous results. This discipline, like all other protective work, was dependent in its nature and intensity upon the struggle for the initiative. One example out of many is found in the numerous German Army Orders which followed our introduction of the Livens projector. This weapon gave the possibility of much higher concentrations at much greater ranges from the front line than were formerly possible and for a time German gas discipline was severely shaken, and the staffs had to react violently to meet the situation. The introduction of this weapon, in fact, was the first clear case of the gaining of the chemical initiative by the Allies. A telegram from German General Headquarters stated: "The English have achieved considerable success by firing gas mines simultaneously from a considerable number of projectors on to one point. Casualties occurred because the gas was fired without warning, and because its concentration was so great that a single breath would incapacitate a man."

This is a further example of the fact that the decisive initiative was very difficult to obtain after two years of war, whereas by the same means it would have been ensured at the commencement. The general development of German protection was a partial safeguard, but the value of the weapon could be seen from the fact that an order was issued for all German working parties to wear gas masks when within 1000 yards of the front line on nights not obviously unsuitable for Allied gas discharges. It is difficult to exaggerate the military importance of such an imposition.

Summary.—We have thus covered a period, the main features of which were attempts at the cloud initiative by Germany and our rapid and successful protective reaction. The conditions surrounding the first attack were entirely peculiar. The complete surprise attending it could only be repeated at the commencement of another war. It failed for entirely different reasons from those which prevented the decisive use of phosgene by the Germans. But our reaction carried us further, and we developed the final form of cloud gas attack, the Livens projector, which, in its turn, taxed the German protection to the utmost, and threatened to overcome it. History repeated itself with a vengeance in this protective struggle.

Two attempts at the cloud initiative, the German phosgene attempt and the Livens projector, were both partially successful. Had either of those attempts shared the surprise of April 22nd, 1915, their success would have been many times greater. It was contact on the battle front that developed a protective appliance and organisation, by giving us an insight into enemy appliances and projects. We cannot emphasise too strongly the significance of this for the future. Apart from remote exceptions, contact will be entirely absent. We can have no guarantee whatever that new devices will be revealed, either between nations or to a central body. Suppose the Germans had been more fully aware of the possibilities of cloud gas, and, realising the dependence of their one method upon wind

134

direction and caprice, had developed our method of producing cloud at a distance. The combinations of the two methods at Ypres could hardly have left a margin of chance for failure. This is a feeble example of what may occur.

New German Attempts.—By this time it was not easy to see how either side could obtain a decisive surprise by the use of chemicals aimed at the respiratory system. It appeared very difficult to penetrate the different forms of respirators by conditions obtainable in the field.

Professor F. P. Kirschbaum, writing on gas warfare, in Schwarte's book, reveals how Germany counted on obtaining the gas initiative against the French at Verdun. He explains how the decision to use Green Cross on a large scale coincided with certain modifications in the design of the German gas shell, which made its large-scale manufacture much simpler and more rapid. "The manufacture of Green Cross," he also tells us, "was assured in the special progress in technical chemistry, and the output was adequate," and goes on to explain, "The first use of per stuff[1] found the enemy unprepared with any suitable protection. The French had equipped their troops with protection against chlorine, but had provided no protection against phosgene,"—"the results of Green Cross ammunition were recognised by the troops. During the big operations before Verdun, however, the enemy did their very utmost to substitute the gas mask M2 for the respirator XTX. Gas mask M2 was a protection against Green Cross. For this reason Green Cross ammunition alone could not be expected to have an effect, as soon as the enemy carried out defensive measures by means of gas mask M2 or

[1] Diphosgene or Green Cross constituents.

some better apparatus. This reverse spurred on the Germans to renewed efforts." The writer proceeds to explain how in 1916 these efforts resulted in finding two important substitutes, mustard gas or Yellow Cross and the arsenic compounds of the Blue Cross type.

Yellow and Blue Cross.—The Germans had, somewhat hastily, laid aside their cloud activities. But they were very keenly pursuing another line, the development of shell gas. Thus, in July, 1917, they made two distinct attempts to regain their initiative by the use of shell gas, and were very largely successful in one case. We refer to the Yellow and Blue Cross shell, containing mustard gas and diphenyl-chlorarsine respectively.

Captain Geyer, writing in Schwarte's book, relates: "Gas was used to a much greater extent, over 100,000 shells to a bombardment after the introduction of the Green Cross shell in the summer of 1916 at Verdun. From that time the use of gas became much more varied as the number of types of guns firing gas projectiles was increased, field guns having also been provided with gas projectiles. The most tremendous advance in the use of gas by the artillery, and indeed in the use of gas in general, came in the summer of 1917 with the introduction of the three elements, Green, Yellow, and Blue, one after another. This introduced the most varied possibilities of employing gas, which were utilised to the full in many places on the front during the successful defensive operations of 1917, above all in Flanders and at Verdun. The hardly perceptible poisoning of an area by means of Yellow Cross shell and the surprise gas attack became two of the new regulation methods of using gas."

Yellow Cross.—The respirator afforded complete protection against the attacks of mustard gas on the respiratory system, but this gas evaded protection in other ways. In the first place, its early unfamiliarity evaded the gas discipline of the Allies, and it was not realised in many cases that the respirator was necessary. This was speedily corrected, but its second line of attack was not easily, and never finally countered. We refer to its vesicant action. Mustard gas could produce severe blistering and skin wounds in such slight concentrations, even through clothing, that it was a tremendous casualty producer, putting men out of action for several weeks or months, with a very low rate of mortality. Used in large quantities against an entirely unprotected army, its results might well have been decisive.

This was the first example of chemical attack upon a new function. We had too readily assumed that gas, or chemical attack, would be restricted to the respiratory system, or to the eyes. We had assumed that if our mask protection was ahead of enemy respiratory attacks our situation was safe. Mustard gas was a rude awakening. It was impossible to protect fully against mustard gas, unless we protected the whole body, and it was never possible to do this during the war without too seriously influencing the movements of the soldier.

Blue Cross.—The Blue Cross Shell was a deliberate attempt to pierce the respirator. It represented to the German mind such an advance of aggression over protection that the effect on the enemy would be almost as if he were entirely unprotected. Some idea of the German estimate of its importance can be found in the following quotation from Captain Geyer: "The search for new irritants in the sphere of arsenic combinations led to the discovery of a series of effective substances. In view of the obvious importance of highly irritant compounds capable of existing in a very finely divided, pulverised, or particulate form, research was made in the domain of little volatile substances with boiling points up to 400°. This led to the astonishing discovery that *diphenylarsenious chloride* when scattered would penetrate all gas masks then in use, even the German, practically unweakened, and would have serious irritant effects on the wearers. This discovery could only be explained by the supposition that the irritant works in the form of particles which it is difficult to keep back by means of a respirator, even a completely protecting respirator, such as the German and English gas masks were at that time. Further analysis showed that the mixture of air and gas examined revealed a concentration of gas greatly in excess of the point of saturation for the vapour given off by this stuff. Finally, ultra microscopic examination showed the existence of smoke particles. A new type of fighting material had been discovered."

He also tells us how, following this discovery, production rose to 600 tons monthly, and used up all the arsenic obtainable in Germany. The Allies were fully alive to the importance of this matter, and we have already explained that, had they been in possession of large quantities of Blue Cross compounds, they might have forced German protection into an impossible position. No better example could be found of the immense superiority enjoyed by Germany owing to her flexible and efficient producing organisation. Captain Geyer goes on to explain how the military value of these projectiles was considerable, and, therefore, the monthly production reached a figure of over one million shell.

We have already emphasised the question of design in chemical warfare, and its importance is borne out by the comparative failure of these German projectiles. Geyer explains how only minute particles less than 1/10,000 of a millimetre in diameter are of any use to penetrate a mask, and he develops the difficulties experienced by Germany in obtaining such fine pulverisation without decomposing the substance. He explains the difficulties which they had in arriving at a suitable shell, and their unsuccessful struggle to overcome the necessity of a glass container, which, he says, demanded "a considerable advance in the technical work of shell production."

This attempt at the chemical initiative by the use of Blue Cross illustrates another method of attack. Geyer says, "Blue and Green Cross ammunition were used simultaneously in the field—called coloured cross (Buntkreuz) in order, by the use of Blue Cross, to force the enemy to remove gas masks, whereby they exposed themselves to the poisonous effects of Green Cross. Matters seldom reached that point, however, for as soon as the enemy realised the effect of 'coloured cross' ammunition, they withdrew troops which were being bombarded with it from their positions to a zone beyond the range of artillery fire. The English in particular had tried to protect the troops against the effects of diphenylarsenious chloride, and of diphenylarsenious cyanide (which followed it and was even more effective) by the use of filters made of woollen material and wadding. They were to a great extent technically successful, but the most effective defensive apparatus, the 'jacket' to the box, was unsatisfactory from the military point of view, as the troops could only make a limited use of it owing to the difficulty of breathing or suffocation which it occasioned."

The Riddle of the Rhine

The reference to the withdrawal of troops is a picturesque misrepresentation. The relative inefficiency of the German shell rendered this unnecessary. In addition, as Captain Geyer explains, our troops were specially protected in anticipation of the use of particulate clouds. An examination of our protective device by the Germans obviously led them to believe that resistance to breathing was too great for the protective appliance to be practicable. But here the exceptional gas discipline of the British troops became effective. There is no doubt that the new mask was worn just as constantly and satisfactorily as the old. Captain Geyer's remarks are also interesting from a point of view to which we have already referred: they show how much this question of resistance to breathing was exercising the minds of those responsible for German protection.

"Particulate" Clouds.—The principle of particulate clouds was not entirely new, both sides having used smoke combined with lethal gases with the object of forcing the removal of the respirator. It was thought that the particulate form of the smoke would penetrate a respirator designed purely to hold up vapours and gases. The reasoning was perfectly sound. It was only a question of using the right smoke in the right way. There were good grounds to believe that such substances would penetrate the respirator, and either produce a casualty or compel the removal of the respirator by the paroxysms produced, to allow some lethal gas to complete the work on the unprotected soldier. Fortunately for us, these objectives were not attained, but this was rather due to some hitch or miscalculation in the German preparations than to any inherent impossibility.

After this period, although chemical warfare

became increasingly an organic part of German (and Allied) operations, yet there is no serious field evidence of a deliberate attempt at the gas initiative. It must be remembered, however, that gas figured very largely indeed in the March, 1918, attempt, by Germany, to regain the general initiative. It is stated authoritatively, for example, that in July, 1918, the German Divisional Ammunition Dump contained normally 50 per cent. of gas shell and, in the preparation, in May, 1918, for German attacks on the Aisne, artillery programmes included as much as 80 per cent. gas shell for certain objectives.

Potential Production and Peace.—Enough has been said to show the general nature of the chemical warfare struggle. The question of the chemical initiative is vital at the commencement of hostilities. Unless, then, we completely rule out any possibility whatever of a future war, it is vital for that occasion. We have indicated sufficiently clearly the factors upon which such initiative depends, to show the critical importance of manufacturing capacity, and protective preparedness.

A further quotation from Schwarte's book is very much to the point. It tells us:

"Whilst on our side only a few gases were introduced, but with successful results, the use of gas by the enemy presents quite another picture. We know of no less than twenty-five gases used by the enemy, and of fifteen types of gas projectile used by the French alone, and we know, from 'blind' (dud) shells which have been found, what they contain. The only effective gases amongst them were phosgene and dichlorodiethyl sulphide. The other substances are harmless preparations, used most probably for purposes of camouflage, a method

141

highly esteemed by the enemy, but which did not enter into the question with us, owing to the capacity of our chemical industry for the production of effective materials."

This is true to a considerable extent. Our dependence on improvised and relatively inefficient production imposed conditions upon Allied policy, whereas, in Germany, they had but to command a flexible and highly efficient producing machine.

The world movement towards disarmament will hardly countenance the maintenance of permanent chemical arsenals. In the face of war experience and further research developments the laborious war improvisation of these arsenals will not save us as it did in the last struggle. Any nation devoid of the means of production invites enemy chemical aggression and is helpless against it. This, and the need to keep abreast of chemical warfare development—particularly in protection—are the chief lessons of the struggle for the chemical initiative.

CHAPTER VII

REVIEW OF PRODUCTION

Critical Importance of Production.—Our analysis
of the struggle for the initiative reveals the critical
importance of production. In the chemical more
than in any other form of warfare, production has a
tactical and strategic importance and functions as
an organic part of the offensive scheme. A tendency
in modern war is to displace the incidence of initiative
towards the rear. Staffs cannot leave the discoveries of the technical workshop or scientific laboratory
out of their calculations, for their sudden introduction into a campaign may have more influence on its
result than the massing of a million men with their
arms and equipment for a surprise assault. The
use of a new war device may shake the opposing
formations more than the most cunningly devised
attack of this sort.

When, after the first brilliant assault on the
Somme on July 1st, we began to lose men, material,
and the initiative, in an endless series of local attacks,
we were even then regaining it by the home development of the tank. Even before the colossal
German effort was frustrated by the first Marne
battle and the development of trench warfare, the
German laboratories were within an ace of regaining
the initiative by their work on cloud gas. After
the lull in their gas attacks, when the Germans
sought to gain the initiative and a decision by the

use of phosgene, the quiet work of our defensive organisations at home had completely countered the move weeks before.

But in all these cases the counter idea could not become effective without large-scale production. This was absolutely fundamental. Had we taken six years to produce the first type of tank, had the Germans failed to manufacture mustard gas within a period of years instead of succeeding in weeks, and had the box respirator taken longer to produce, all the brilliant thinking and research underlying these developments would have had practically no influence on the campaign, for they would have had no incidence upon it. We could go on multiplying examples. But what is the conclusion?

From this rapid development of methods a new principle emerges. The initiative no longer remains the sole property of the staffs unless we enlarge the staff conception. Vital moves can be engineered from a point very remote in organisation and distance from the G.H.Q. of armies in the field. But there is a critical step between the invention and its effect on military initiative. This is production, which for these newer methods becomes an organic part of the campaign.

But the future is our chief preoccupation. What would be the supreme characteristics of the early stages of a future war? It would be distinguished by attempts of belligerents to win immediate and decisive success by large scale use of various types of surprise. Three factors would be pre-eminent, the nature of the idea or invention, the magnitude on which it is employed, and its actual time of incidence, that is, the delay between the actual declaration of war and its use. Now the invention is of no use whatever without the last two factors, which are entirely dependent on production. When, in 1917, the Allied staffs pressed repeatedly for gases with which to reply to German Yellow Cross, their urgent representations met with no satisfactory response until nearly a year had elapsed. This was not due to lack of invention, for we had simply to copy the German discovery. Failure to meet the crying demands of the Front was due to delay in production.

Any eventual chemical surprise will, under genuine conditions of disarmament, depend on peace industry, for no such conditions will tolerate the existence of huge military arsenals. We have already indicated the type of peace-time industry *par excellence*, which can rapidly and silently mobilise for war. It is the organic chemical industry. Therefore, whatever the war may have taught us as to the value of chemical industry, its importance from the point of view of a future war is magnified many times. The surprise factor is responsible. The next war will only commence once, however long it may drag on, and it is to the start that all efforts of a nation planning war will be directed. It is, therefore, of importance to examine in detail the development of chemical production during the recent war.

A close examination is of more than historical significance, and should provide answers to certain vital questions. German chemical industry was the critical factor in this new method of war which almost led to our downfall. How did the activities of this industry compare with our own production? To this an answer is attempted below, but graver questions follow. Was our inferior position due to more than a combination of normal economic conditions, and were we the victims of a considered policy? If so, who directed it, and when did it first give evidence of activity? An answer to these questions will be attempted in a later chapter.

Significance of the German Dye Industry.—At the end of 1914 the nation began to realise what it meant to be at the mercy of the German dye monopoly. Apart from the immediate economic war disadvantages, the variety and sinister peace ramifications of this monopoly had not been clearly revealed. Mr. Runciman, then President of the Board of Trade, stated with regard to the dye industry: "The inquiries of the Government have led them to the conclusion that the excessive dependence of this country on a single foreign country for materials of such vital importance to the industry in which millions of our workpeople were employed, constitutes a permanent danger which can only be remedied by a combined national effort on a scale which requires and justifies an exceptional measure of State encouragement." Measures were defined later.

In the debate in the House of Commons in February, 1915, on the aniline dye industry, a member prominent in the discussion, referring to "taking sides on the question of Free Trade," stated that, "It was a great pity that this should occur when the attention of the House is occupied with regard to *matters connected with the war*," and proceeded to draw a comparison between the national importance of the manufacture of dyes and that of lead pencils. Fortunately he prefaced his remarks by explaining his ignorance of the "technical matters involved in this aniline dye industry." These are two out of many references to the pressure due to the absence of German dyes, which illustrate the purely economic grounds on which the issue was being discussed, on

146

the one hand, and reveal the prevailing ignorance of its importance on the other.

Exactly one month later came the first German gas shock. Such statements as the above tempt us to ask who, at this time, realised the common source of the direct military and indirect economic attack. It can hardly be doubted that the existence of the German dye factories was largely responsible for the first German use of gas on the front. We have already seen how, from the first month of the war, the chemical weapon was the subject of definite research. Falkenhayn leaves us in no doubt as to the chief factor which finally determined its use. Referring to difficulties of production, he says, "Only those who held responsible posts in the German G.H.Q. in the winter of 1914-15 . . . can form any estimate of the difficulty which had to be overcome at that time. The adjustment of science and engineering . . . took place almost noiselessly, so that they were accomplished before the enemy quite knew what was happening. Particular stress was laid upon the promotion of the production of munitions . . . as well as the development of gas as a means of warfare." Referring to protective methods of trench warfare, he continues, "Where one party had gained time . . . the ordinary methods of attack often failed completely. A weapon had, therefore, to be found which was superior to them but which would not excessively tax the limited capacity of German war industry in its production. Such a weapon existed in gas."

The Germans had themselves shown us where this production occurred, and Ludendorff supplements our information by telling us how he discussed the supply of war material with Herr Duisburg and Herr Krupp von Bohlen in Halbach, "whom I had asked to join the train" in the autumn of 1916. The former was the Chairman of the I.G., the great dye combine.

Those producing a new weapon of war must always consider the possibilities possessed by their opponents to exploit the same weapon after the first shock. For the Germans the answer was obvious. The Allies would be held at a material disadvantage for months, if not years. Without the means of production available in Germany, we are not at all convinced that the gas experiment would have been made, and had it not been made, and its formidable success revealed, Germany's hesitation to use this new weapon would probably have carried the day. This, at least, is the most generous point of view. In other words, the German poison gas experiment owed a large part of its initial momentum to ease of production by a monopoly. The combination of this factor with the willingness to use gas led to the great experiment. The future may again provide this combination, unless the monopoly is removed.

Following up this line of thought, we can see how tempting was the German course of action. Falkenhayn has told us what a violent strain was imposed upon Germany by the stabilisation of the Western Front early in 1915. The tension between the Great General Headquarters and the Home Government was already in evidence, and would have caused difficulty in attaining suitable home and liaison organisations, in particular with regard to supply. We can well understand this when we remember the drastic changes which occurred in our own ministries and departments. But what organisation was required for chemical warfare supply? Very little! Quoting from the report of the Hartley

Mission to the chemical factories in the occupied zone, we know that when the Government wished to produce a new gas "a conference with the various firms was held at Berlin to determine how manufacture should be subdivided in order to use the existing plant to the best advantage." The firms referred to were the constituent members of the highly organised I.G. There was no need to create a clumsy and complicated organisation with an efficient one existing in the I.G. ready to meet the Government demands. The path could not have been smoother. Ludendorff states in his memoirs that the Hindenburg programme made a special feature of gas production. Increased supply of explosives was also provided for. He says: "We aimed at approximately doubling the previous production." And again: "Gas production, too, had to keep pace with the increased output of ammunition. The discharge of gas from cylinders was used less and less. The use of gas shells increased correspondingly." This programme represented a determined effort to speed up munitions production in the autumn of 1916. It included not only gas but explosives, both of which could be supplied by the I.G. Explosives demanded oleum, nitric acid, and nitrating plants, which already existed, standardised, in the factories of the dye combine. The unusual speed with which standard dye-producing plant was converted for the production of explosives is instanced in the operation of a T.N.T. plant at Leverkusen, producing 250 tons per month. The conversion only took six weeks. The factories of the I.G. supplied a considerable proportion of the high explosives used by Germany.

In the field of chemical warfare the relationship between war and peace production was even more

intimate. Chemical warfare products are closely
allied and in some cases almost identical with the
finished organic chemicals and intermediates produced by the dye industry. Therefore, in most
cases, even when the suggestion of the new chemical
may come from a research organisation entirely
apart from the dye research laboratories, the products fall automatically into the class handled by
the dye industry.

Is there any doubt that the I.G. was a terribly
effective arsenal for the mass production of the
older war chemicals, explosives, and the newer types,
poison gases? Is there even a shadow of exaggeration in our claims? There may be those who
would see a speedy resumption of friendship with
Germany at all costs, regardless of the honourable
settling of her debts, regardless of her disarmament
and due reparation for wrongs committed. Can
even such concoct material to whitewash the military front of the I.G.? If they would, they must
explain away these facts.

The plants of the I.G. produced more than two
thousand tons of explosives per week, at their
average pre-war rate. This is an enormous quantity. How can we best visualise it? In view of the
chapters on Disarmament which follow, we will
use the following comparison. The Treaty of Versailles allows Germany to hold a stock of about
half a million shell of different stated calibres. How
much explosive will these shell require? They
could be filled by less than two days' explosives production of the I.G. at its average war rate. Between
two and three million shell could be filled by the
result of a week's production in this organisation.
Further, the average rate of poison gas production
within the I.G. was at least three thousand tons per

month, sufficient to fill more than two million shell of Treaty calibres. Unless drastic action has been taken, the bulk of this capacity will remain, and Germany will be able to produce enough poison gas in a week to fill the Treaty stock of shell; this in a country where the manufacture and use of such substances are specially prohibited.

It is appropriate at this stage to describe as briefly as possible the origin and composition of this great German combination, the Interessen Gemeinschaft, known as the I.G. There is no need to go into the gradual self-neglect, and the eventual rooting out by Germany, of the dye-producing industry in other countries, notably England, France, and America.

The Interessen Gemeinschaft.—By the end of the nineteenth century the manufacture of dyes on a large scale was concentrated almost exclusively in six great firms. These were the Badische Anilin und Soda Fabrik, Ludwigshafen on the Rhine, known as the Badische; the Farbenfabriken vorm. Friedr. Bayer & Co., in Leverkusen, known as Bayer; Aktien-Gesellschaft für Anilin-Fabrikation in Berlin; Farbwerke vorm. Meister Lucius & Brüning in Höchst am Main, referred to as Höchst; Leopold Cassella G.m.b.H. in Frankfort; and Kalle & Co., Aktien-Gesellschaft in Biebrich.

Each of these six great companies had attained enormous proportions long before the war. Only two other concerns had carried on manufacture on a comparable scale. These were the Chemische Fabrik Greisheim-Elektron of Frankfort A.M., a company which has absorbed a number of smaller manufacturers, and the Chemische Fabriken vormals Weiler-ter Meer, Uerdingen.

The position of all these establishments, with

one single exception, along the Rhine and its tributaries is well known. Their growth has been illustrated in their own prospectuses. Höchst was organised in 1863 and started with five workmen. In 1912 it employed 7680 workmen, 374 foremen, 307 academically trained chemists, and 74 highly qualified engineers. The works of the Badische, which was organised in 1865, covered, in 1914, 500 acres, with a water front of a mile and half on the Rhine. There were 100 acres of buildings, 11,000 workmen, and the company was capitalised at fifty-four million marks. The establishment of Bayer was on a scale entirely comparable. Quoting from an official American report,[1] "Griesheim Elektron, prior to the war, had enormous works chiefly devoted to the manufacture of electrolytic chemicals and became an important factor in the dyestuff business only within recent years, when by absorption of the Oehler Works and the Chemikalien Werke Griesheim, its colour production reached a scale approaching that of the larger houses." This move on the part of the Griesheim Elektron is interesting as an example of the general tendency which has characterised the development of the German dye industry. This firm, producing inorganic materials and intermediates, absorbed the Oehler Works in order to find an independent outlet for its intermediate products, thus becoming directly interested in dyestuffs production. This move towards independence in the whole range of products involved is referred to elsewhere, owing to the manner in which it simplified German production for chemical warfare.

Combination, however, did not cease in the creation of these enormous establishments. The cartel

[1] Alien Property Custodian's Report, 1919.

fever raged here as in other German industries. By 1904 two immense combinations had been formed in the dyestuff industry. One of these comprised Bayer, Badische, and Berlin; the other Höchst, Cassella, and Kalle. "By pooling profits, by so arranging capitalisation that each company held stock in the other companies of its own cartel, and by other familiar means, the risks incident to the enormous expansion of the business and the immense increases of export trade were minimised. The centripetal tendency, however, did not stop here. In 1916, the two pre-existing cartels were combined with Griesheim Elektron, Weilerter Meer, and various smaller companies in one gigantic cartel, representing a nationalisation of the entire German dye and pharmaceutical industry." The combination was extremely close. Profits of the companies were pooled, and after being ascertained each year on common principles were divided according to agreed percentages. Each factory maintained an independent administration, but they kept each other informed as to processes and experiences. "There was also an agreement that in order to circumvent tariff obstacles in other countries materials were to be produced outside of Germany by common action and at common expense whenever and wherever desirable.

"At the time of the formation of this enormous organisation the capitalisation of each of the principal component companies was largely increased. Höchst, Badische, and Bayer each increased their capitalisation by 36,000,000 marks, bringing the capital of each up to 90,000,000 marks." "Berlin increased its capital from 19,800,000 to 33,000,000 marks. Other increases brought the total nominal capital of the group to over 383,000,000 marks. For

153

many years a large part of the enormous profits of these concerns has been put back into the works with the result indicated by the stock quotations. The real capitalisation is thus much greater than this nominal figure. In fact, it is estimated that the actual investment in the works comprising the cartel is not less than $400,000,000. It cannot be doubted that this enormous engine of commercial warfare has been created expressly for the expected war after the war, and that it is intended to undertake still more efficiently and on a larger scale the various methods by which German attacks upon all competition were carried on."

Two additional features must be indicated. A policy to which we have referred was most actively followed, aiming at complete independence and self-sufficiency in all matters relevant to production, especially regarding raw materials. We mention later how the war has strengthened the strong pre-war position of the I.G. in heavy chemicals needed as raw materials for the intermediates and finished dyes.

Recent information reveals a further widening of their basis of operation, including a strong hold on the electro-chemical industry and on the new synthetic processes from carbide, for acetic acid and the other products normally obtained by wood distillation. Again, the policy of the I.G. appears to have moved towards more complete unity since the war. Exchanges of directing personnel and of capital amongst the branches have been recorded for which the term "cartel" is no longer a fair description. In addition, considerable increases in capital have occurred which not only reveal the vision and activity of the I.G. but which indicate its close contact with the German Government. With such an organisation in existence and with the complete liaison which had developed between the directors and the German Government for other purposes than chemical warfare, and in agreement with the paternal policy adopted by the latter towards this chemical industry, production became simplicity itself.

War Production by the I.G.—Let us, therefore, examine in some detail the actual production of war gases and chemicals by the I.G. In order to obtain an idea regarding ease of production, we will later make a comparison with the magnitude and rapidity of that of the Allies.

From the point of view of this statement, there are two main classes of production, that in which the majority of the steps involved were actual processes employed for the manufacture of some dye, pharmaceutical or other chemical product, and, in the second place, that in which no such coincidence occurred, but in which the general technique developed, and the varieties of existing plant covered the needs of the case. Without stretching the point, every war chemical employed came easily under one of these two categories. In order to assist the less technical reader, we will consider the production of the chief war chemicals in the order in which they appeared against us on the front.

Chlorine.—This important raw material, used in a variety of operations, notably for the production of indigo and sulphur black, two highly important dyes, was produced along the Rhine before the war to the extent of nearly forty tons a day. The only serious expansion required for war was an increase of already existing plant at the large factory of Ludwigshaven. The following table of production illustrates the point:

CHLORINE (METRIC TONS PER DAY)

	1914	1918
Leverkusen	20	20
Höchst	4	8
Ludwigshafen	13	35
Total	37	63

Chlorine was important, not only as a raw material for most of the known chemical warfare products, but also, in the liquid form, for cloud attack. Owing to the development of protection, the use of liquid chlorine for the latter purpose became obsolete.

Phosgene.—This was produced in considerable quantity before the war at Leverkusen and Ludwigshafen, leading to many exceedingly important dyes, amongst the most commonly used at present being the brilliant acid fast cotton scarlets so largely used in England. More expansion of plant was necessitated. At Leverkusen the existing plant can produce at least thirty tons a month, and we learn "the plant remains intact ready for use." At Ludwigshafen the capacity was considerably higher, amounting to 600 tons per month. As production was commenced before the war, there were no difficulties in developing the process, expansion alone being necessary.

Xylyl Bromide.—This was one of the early lachrymators, and was produced at Leverkusen in a plant with a maximum monthly output of sixty tons. Production began, according to a statement on the works, in March, 1915. Its ease can be judged from the fact that this compound was used almost as soon as the first chlorine cloud attack at Ypres.

The Germans undoubtedly attached considerable importance to their brominated lachrymators. In this connection their persistent efforts to retain the bromine monopoly with their Stassfurt product and to crush the American industry before the war are significant. The success of these efforts certainly placed us in a difficult situation during the war, both with regard to production of drugs and lachrymators.

German bromine was associated with potash in the Stassfurt mineral deposits, whereas the American product was produced from numerous salt springs and rock salt mines. Although Germany had not succeeded in crushing the American industry, yet the outbreak of war found her in a predominant position, for her two chief opponents, France and England, were cut off from their supplies, which were German; and American production was of little use, owing to the great excess of demand over supply, and the manipulation of output by German agents in America. A possible source of bromine existed in the French Tunisian salt lagoons, whose pre-war exploitation had been considered by an Austrian combination. The French wisely developed a Tunisian bromine industry sufficient for their own needs, and, on different occasions, supplied us with small quantities. But the development of such an enterprise in time of war was a severe handicap.

Diphosgene or Trichlormethyl Chloroformate.— This substance was toxic, a lachrymator, and slightly persistent. It attained a maximum monthly output of 300 tons at Leverkusen, and about 250 tons at Höchst. This was not a simple compound to make, and had no direct relationship with the stable product of the peace-time industry. At the same time, it provides an example of the way in which general technique developed by the industry was

157

rapidly used to master the new process. In particular their method of lining reaction vessels was of value here. The reaction occurs in two stages by the production of methyl formate and its subsequent chlorination. The methyl-formate plant was part of an existing installation, but the chlorination and distillation plant were specially installed.

Chlorpicrin.—This was mixed with diphosgene and used in the familiar Green Cross shell. The production was very readily mastered and attained the rate of 200 tons per month. Picric acid, chlorine, and lime were required, all three being normal raw materials or products of the industry. At Höchst no new plant was installed, the manufacture being carried out in the synthetic indigo plant.

Phenylcarbylamine Chloride.—This was used in German chemical shell, and was not particularly effective against us, although produced in large quantities by the Germans, in vessels used in peace time for a very common intermediate, monochlorbenzene. The ease of production of this substance may account for its use in large quantities by the Germans, in order to increase their gas shell programme.

Mustard Gas or Dichlordiethyl Sulphide.—This was prepared in four stages:

(1) Preparation of Ethylene—by heating alcohol with an aluminium oxide catalyst at 400° C.
(2) Preparation of Ethylene-chlor-hydrin, by passing ethylene and carbon dioxide into a 10 per cent. solution of bleaching powder at a temperature below zero centigrade, and subsequent concentration of the product to a 20 per cent. solution.
(3) Conversion of the chlor-hydrin into thio-

158

diglycol by treatment with sodium sulphide.

(4) Conversion of the thiodiglycol into mustard gas (dichlordiethyl-sulphide), using gaseous hydrochloric acid.

The thiodiglycol was produced at Ludwigshafen and provides one of the best examples of the adaptation of the German dye works for the purpose of producing war chemical. Technically, ethylene is a fairly difficult gas to produce in large quantities, but, for the Ludwigshafen works, these difficulties were a thing of the past. There were twelve big units before the war, and, by the time of the Armistice, these had been increased to seventy-two in connection with mustard gas manufacture. In a similar way, the number of the units for chlorhydrin, the next step, was increased from three to eighteen. These two processes had all been worked out very thoroughly in connection with the production of indigo. These new plants were identical with the peace-time units. The expansion was a mere question of repetition requiring no new designs or experiments and risking no failure or delay. Success was assured. The last step, the production of thiodiglycol, occurred in the causticising house, to which no substantial alterations or additions appear to have been made for the purpose. As sodium sulphide is used in large quantities as a raw material in the dye industry, and was already produced within the I.G., no difficulty was presented in connection with its supply.

The thiodiglycol was forwarded to two other factories, one of which was Leverkusen, where 300 tons of mustard gas were produced monthly. The reaction between thiodiglycol and hydrochloric acid was one which required very considerable care.

159

At one stage of the war the Allies viewed with much misgiving the possibility of having to adopt this method. But the technique of the German dye industry solved this as satisfactorily and as steadily as other chemical warfare problems, bringing its technical experience to bear on the different difficulties involved.

Diphenychlorarsine.—This was the earliest and main constituent of the familiar Blue Cross shell. It was prepared in four stages:

(1) The preparation of phenyl arsinic acid.
(2) The conversion of the above to phenyl arsenious oxide.
(3) The conversion of the latter into diphenyl arsinic acid.
(4) The conversion of the latter into diphenyl-chlor-arsine.

This is another example of a highly complicated product which might have presented great difficulties of production, but the problem of whose manufacture was solved, almost automatically, by the German organisation.

The first step, that of the manufacture of phenyl arsinic acid, was carried out at Ludwigshafen in one of the existing azo dye sheds without any alteration of plant, just as a new azo dye might have been produced in the same shed. It is believed that another dye factory also produced this substance. At Ludwigshafen the conversion to diphenyl arsinic acid occurred. This was again carried out in the azo colour shed, with no more modification than that involved in passing from one azo dye to another.

This chemical mobilisation of a huge dye unit was, and could still be, practically invisible in operation. Not only was the process practically the same as azo dye production, but, as the compounds were not particularly poisonous in the intermediate stages, there was no risk to the workers, and no need to violate secrecy by indicating special precautions.

The final stage, the preparation of diphenyl-chlorarsine, the actual Blue Cross shell constituent, occurred at Höchst, which also carried out the first three stages, already outlined as occurring at Ludwigshafen and Leverkusen. The last stage was a simple one and was carried out in plant and buildings previously used for peace purposes.

The other substances employed provide further examples of this ease of production. Ethyl-dichlorarsine was produced in homogeneously lead-lined vessels, identical with those used for diphosgene. Dichlor-methyl-ether presented difficulties which were solved by applying the German method of using tiled vessels.

The part played by the I.G. in the German chemical warfare organisation has already been outlined, and we have seen how the German Government was content simply to place its demands before the directors of the dye combine. The latter were left to choose the process and exploit it by making the best use of their organisation, which was done after reviewing the plant at their disposal in the different branches. An interesting feature of the production of war chemicals by the I.G. is thus revealed by examining the actual locality of the separate operations leading to any one of the individual poison gases. The attached table shows us how the production of any particular war chemical involved a number of stages, each of which occurred in a different factory. The directors of the I.G. simply chose a particular plant in a particular factory which was most suited for the operation concerned. They

WAR CHEMICAL	RAW MATERIALS FROM THE I.G.	FIRST STAGE	
		PROCESS	FACTORY
Phenyl Carbylamine Chloride	1. Aniline 2. Chlorine 3. Caustic soda	Condensation of *aniline* with *carbon bisulphide* to *phenyldithiocarbamic acid*	Kalle
Mustard Gas	1. Carbon dioxide 2. Bleaching powder 3. Sodium sulphide 4. Hydrochloric acid	Preparation of *Ethylene* from *Alcohol*	Ludwigshafen
Diphenylchlorarsine	1. Aniline 2. Sodium nitrite 3. Sodium bisulphite 4. Sodium hydrate 5. Sulphur dioxide 6. Hydrochloric acid	Conversion of *Diazobenzene* to *Phenylarsinic acid*	Ludwigshafen Kalle Höchst
Ethyl-dichlorarsine	1. Ethyl chloride 2. Caustic soda 3. Sulphur dioxide 4. Hydrochloric acid gas 5. Iodine	Production of *Ethylarsinic acid* from *Ethyl chloride*	Ludwigshafen
Sym-dichlor-methyl-ether	1. Chlorsulphonic acid 2. Sulphuric acid	Production of *Formaldehyde* from *Methyl alcohol*	Mainz Höchst

SECOND STAGE		THIRD STAGE		FOURTH STAGE	
PROCESS	FACTORY	PROCESS	FACTORY	PROCESS	FACTORY
nversion of *Phenyldithio-arbamic acid* o *Phenyl Mus-ard Oil* by :inc chloride	Kalle	Chlorination of *Phenyl Mus-tard Oil* giving *Phenyl Carby-lamine Chlo-ride*	Höchst		
nversion of *Ethylene* into *Ethylene Chlorhydrin*	Lud-wigs-hafen	Conversion of *Chlorhydrin* to *Thio-di-glycol*	Lud-wigs-hafen	Conversion of *Thiodiglycol* to *Mustard Gas*	Lever-kusen
duction of *Phenyl arsinic acid* to *Phenyl arsenious oxide*	Lever-kusen and Höchst	Conversion to *Diphenylar-sinic acid* by treatment with *Diazo benzene*	Lever-kusen and Höchst	Reduction of *Diphenylar-sinic acid* to *Diphenyl-chlor-arsine* by *Sulphur dioxide* in HCl solution	A.G.F.A. Höchst
duction of *Ethyl arsinic acid* to *Ethyl arsenious oxide* by *sulphur dioxide*	Lud-wigs-hafen	Conversion of *Ethyl arseni-ous Oxide* to *Ethyl dichlor-arsine* by HCl and iodine	Höchst		
nversion of *paraformalde-hyde* to *sym dichlor methyl ether* by means of *chlorsul-phonic acid*	Höchst				

aimed at the minimum conversion, and in a number of cases none was required. The above analysis can leave us with no doubt in our minds that the organic chemical industry is the logical place for efficient chemical warfare production. It cannot leave us unconvinced as to the vital importance of the dye industry in national defence.

Allied Difficulties.—Our own production was nothing but a series of slow and relatively inefficient improvisations. We have already referred to the fluctuations in chemical warfare organisation for research and supply during the war. These added to the difficulties of the supply department, just as they did to its complement, the research department. Only great patriotic endeavour could have made possible the relative success achieved, not only by the departments, but in particular by the firms with whom they were called upon to co-ordinate.

We wanted mustard gas, and realised its need in July, 1917. Research work began almost from that date, yet successful large scale production did not materialise in England until more than a year later. We must admit, however, that the French were in a position to use their product on the front in July, 1918. Let us examine some of our difficulties.

The first efforts were directed towards the process by which, as we eventually ascertained, the Germans produced the whole of their mustard gas. The actual chemical laboratory details of the process presented no serious obstacle, but difficulties multiplied as soon as we attempted large scale work. We wanted ethylene-monochlor-hydrin. Some work had been done on this during the war for the National Health Insurance Commissioners in connection with the production of novocain. Half scale work had occurred at the works of a Midland chemical firm, and experience so gained was freely offered and used in a scheme for the large scale production of mustard gas by the co-operation of a number of big chemical manufacturers. Pressing requests for the material were continually coming from G.H.Q., the programmes outlined being more and more ambitious. We had to reproduce the result of years of German effort spent in developing their monochlorhydrin process for indigo. As a consequence, large sums of money were expended on the process, although it never eventually operated. Its difficulties, and other reasons, led us to research on other and more direct methods which the French were also investigating. The successful outcome of this early research was due, in particular, to Sir William Pope and those associated with him in the work. The process was so promising that the long and cumbersome chlor-hydrin method was abandoned. As a result our five or six months' work on the German method meant so much time lost. The new direct, sulphur monochloride method was taken up actively and several private firms attempted to develop the small scale manufacture. The work was dangerous. Lack of that highly developed organic chemical technique, which was practically a German monopoly, rendered the task much more dangerous than it would have been if undertaken by one of the I.G. factories.

The French, realising the importance of the new methods, spared nothing in their attempts to develop them. Their casualties multiplied at the works, but the only reply was to put the factories concerned under the same régime as the front, and the staffs were strengthened by well-chosen military personnel. The French realised the nature of their task, and organised for it. When the difficulties of production were pointed out in August, 1917, in the British Ministry of Munitions, reports were instanced that the Germans had used forced labour. The French in their production at Rousillon, on the Rhone, employed volunteer German prisoners. It was a curious contrast to see mingling together amongst the producing plants representatives of the American, Italian, and British Missions, with French officers, French technical men, and German prisoners. The latter appeared to be perfectly satisfied in their work. They were used for certain limited purposes, such as handling raw materials, and were not, as a rule, exposed to the dangerous operations against which the French struggled so heroically and successfully. It was as though a small section of the front had been transferred to the heart of France. We saw the minister visiting a factory and pinning the Legion of Honour on to the breast of a worker blinded by yperite. We saw messages of congratulation from the front to the factories themselves. The morale was wonderful. As a result, the French mastered the technical difficulties of mustard gas production and shell filling by June, 1918. They shared information with us, but the race had started neck and neck, and it was impossible to discard completely the large plants to which we were already committed. Without disparaging our own efforts, we must pay a tribute to the achievement of the French yperite producing and filling factories. It is impossible to give personal credit in this matter without going beyond our scope, and we can only draw general comparisons. But we must draw attention to the following. The German factories passed with ease to mustard gas production by a process which, compared with the final Allied method, was clumsy and complicated, but which suited their pre-war plant. Their policy was, therefore, sound from the point of view of the campaign. The Allies experienced great difficulty and danger in attaining large scale manufacture with a simpler process.

The same self-sacrificing zeal and patriotic endeavour was shown in this country, but we were handicapped in mustard gas production by the energetic way in which we had pressed forward the industrial realisation of the monochlor-hydrin method. The French, less committed in terms of plant and finance, could more readily adjust their energy, materials, and money to the new method. It must not be forgotten, also, that, at this period, chemical warfare supply organisation was experiencing certain critical changes which could not but reflect upon our efficiency. Here again the earlier centralisation of research and production by France was a great factor in her favour.

Our difficulties with phosgene, and in particular with the arsenic compounds described above, were of the same nature, involving us in casualties, great expenditure, and little success, when compared with German production. The great need for these arsenic compounds was realised as early as February, 1918, and investigations began even at that date, but they had not appeared in the field by the time of the Armistice. Whatever mistakes we may have made locally during the war, they are small compared with the big mistake which was responsible for our comparative failure in chemical warfare production. We were almost completely lacking in organic chemical industrial experience.

It is interesting to note that the activities of those elements of organic chemical industry which did exist in France and England fully justified the conclusions we have drawn. Thus, although entering late into the field of chemical warfare production, Doctor Herbert Levinstein, Professor A. G. Green, and their collaborators of the firm of Levinstein Limited were able to develop rapidly a successful industrial mustard gas process which was of considerable assistance to England and America. This work, both in research and production, deserves the greatest credit. Again, the dye factories were called upon much earlier to assist in French production and were of considerable assistance.

It would be well at this juncture to review very briefly the other war activities of our own dye industries. The outbreak of war found them by no means inactive. In this country, for example, our own dye factories were able to keep pace with the increasing demand for dyes created by the rapid mobilisation of military and naval equipment. In particular the rapid large-scale production of indigo by the Levinstein firm, at Ellesmere Port, was a considerable achievement. In addition, the new State-aided enterprise at Huddersfield was largely diverted to explosives production, and rendered very valuable services in the supply of Tetryl, T.N.T., synthetic phenol, picric acid, and oleum. For such reasons, the need for essential dyes, and the use of dye capacity for explosives, the important part which the rapidly expanding industry could have played in chemical warfare production was not recognised quickly enough by the relevant authorities. This is not surprising, for the war significance of the German dye industry was not fully realised until the Armistice. It required the Hartley Mission to drive this fact home. When, however, the brilliant researches, referred to above, on the mustard gas method had decided our policy, the dye factory of Levinstein Limited vigorously converted the process into a technical success, and what was still a laboratory reaction in the spring of 1917 became a successful manufacturing process in July of that year.

Released from its war responsibilities at the time of the Armistice, the British industry developed so rapidly that Lord Moulton, in a speech to the Colour Users Association on November 28th, 1919, stated: "A few months before the war broke out England produced only one-tenth of the dyes she needed, but the amount which I am informed we shall be able to turn out at the end of this year would, in weight, be within one-fifth of the amount which England used before the war."

But the Allies were not only in difficulties with regard to the lack of suitable peace-time plant, and industrial organic chemical experience—they were hindered at almost every turn by difficulties with regard to raw materials and intermediates, the products of other chemical manufacture. They had to create a liquid chlorine industry. In April, 1915, the only liquid chlorine plant in England was in the hands of the firm of Castner Kellner, whose maximum output was not more than a few tons per day. Increase in capacity was rendered necessary by chemical warfare developments. Chlorine was a raw material for mustard gas and practically every important substance employed in chemical warfare including bleaching powder. Tremendous tonnages of bleach were involved in the manufacture of chlorpicrin and for use as an antidote against mustard gas on the front. We refer elsewhere to the developing use of bleach in order to create lanes for troops and transport through areas infected by mustard

169

gas. A very simple calculation will show what quantities would be required for such an operation. It is true that, as regards chlorine, we were more favourably situated than France, and forwarded her considerable supplies in exchange for phosgene. This chlorine was essential for phosgene production. New plants were brought into being at different places, largely through the energy and experience of the above-mentioned firm, but so great was the demand that it finally became necessary, in order to protect the trade users and war interests at the same time, to institute a control of chlorine. More than 20,000 tons of liquid chlorine were produced under the administration of the supply department concerned. When we consider the effort which such an increase in production must have involved, and the fact that expansions occurring did not do so under the steady and well-regulated influence of a simple demand, but were continually being modified to meet expansions or diminutions of programme, we can realise what a great advantage was possessed by the Germans owing to their large initial experience and production.

We have no hesitation in stating that great credit is due to the old Trench Warfare Supply Department and the firms with which it was in contact, notably the one referred to above, in connection with the Loos attack. But for them, we would not have been in a position to retaliate, even at that date.

The Allied lachrymator campaign was terribly handicapped by lack of bromine. The French performed the phenomenal task of creating a bromine industry in Tunis, the development of which reads like a romance. Apparently this industry is dying out, and German predominance in bromine is again asserted.

French mustard gas production, for which they made such huge sacrifices, was threatened by the lack of carbon-tetra-chloride, and examples can be multiplied. The Germans were in a very different position. The development of their dye industry had followed the policy of absolute independence of external chemical industry. This independence was acquired either by the absorption of other enterprises or by the definite development of processes and plant for raw materials and intermediates. In every case the war has strengthened these factories for the manufacture of these products. In 1918 they produced nearly thirty times as much ammonia as in 1914, three times as much nitric acid, fifty per cent. as much again of sulphuric acid, and twice as much liquid chlorine. This was not purely a commercial question. Our lack of such products was due to the fact that the Allies, in pre-war times, possessed few or feeble industries whose consumption would stimulate the production of these raw materials. They lacked these industries because of a blameworthy disregard for the fundamental importance of science, and particularly chemical science, in industry.

Conclusion.—We have shown how, during the war, chemical warfare proved its surprise value and how manufacture figured repeatedly as a critical factor. We have also shown how the importance of production is magnified from the point of view of the future. The only logical conclusion is that the country which does not possess a strong dye industry, or enormously comprehensive and expensive chemical arsenals, cannot hope to escape serious military results, possibly defeat, from enemy chemical surprises. The situation is aggravated by the fact that this critical producing capacity exists as a monopoly in the hands of Germany. No patriotic and thinking person can, therefore, conclude otherwise than to encourage the creation of dye industries in countries other than Germany, particularly in our own. It is true, however, that patriotic sentiment and political views do not always lead to the same solution. But we must insist that there can be no two opinions on the national defence aspect of this question, and any political forces opposing the logical outcome of patriotic sentiment in this case are incurring an exceedingly grave responsibility.

Further, there is a definite tendency to obscure the whole issue by inaccurate thinking. When we find a Member of Parliament seriously discussing disarmament, endeavouring to deal with the matter in detail, and yet classing gas as one of those methods of warfare in connection with which production can be easily prevented,[1] we can only stand in amazement before our traditional fault, deliberate sidetracking of expert guidance. When we realise that it was not until after the Armistice that the Hartley Commission opened our eyes to the war importance of the German dye industry, we see how blind a nation may be in matters vital to its defence.

From the point of view of results on the front, for which all were working, the German dye factories, when considered as a war weapon, were as much in advance of Allied improvised plants as a military quick-firing gun is ahead of the old muzzle-loader.

Further, for progressive and flexible organic chemical production, some such difference will always exist between the modern dye industry and factories or arsenals improvised or maintained to meet specific emergencies.

[1] *The Flaw in the Covenant and the Remedy,* Major David Davies, M.P.

172

CHAPTER VIII

AMERICAN DEVELOPMENTS

Special Attention Justified;—Special Value of American Opinion.—Various reasons prompt us to pay special attention to the development of chemical warfare by the United States of America. In the preceding chapters we have attempted a more or less connected account of its development during the campaign. Such an account must necessarily make constant reference to French and British developments. But American preparations, although on a colossal scale, were not in time to influence the campaign seriously and directly. Therefore, purely for the symmetry of our account, special reference should be made to America. But a more serious reason is to be found in the great importance attached by America to this branch of warfare. As everybody knows, the arrival of the American troops in large numbers was preceded by an educational period, during which American staffs, officers, and men became acquainted with Allied staffs, operations, and methods on the Western Front. They were less biased by military tradition, and not under the same necessity as the European Allies to organise in an improvised way for different violent emergencies. Their opinions of war methods on the Western Front are, therefore, of great interest.

Chemical warfare at once assumed a place of prime importance in their schemes, receiving a stimu-

173

lus and a momentum which, rather than losing force during peace, appears to have gathered intensity. There was at first no particular background of emotion, or desire for specific retaliation in this American development. It was purely a question of deciding on technical grounds the relative importance of different methods of warfare. Solid facts determined the matter later. We have it on the best authority that 75,000 out of the total 275,000 American casualties were due to gas.

Early American Activities.—The earliest American activities consisted in attaching various officers to the British formations in France and to the French research and producing organisations centred in Paris. A period ensued of remarkably rapid and efficient assimilation of the best developments in allied chemical warfare. Two American gas companies were attached to ours for instruction in the first month of 1918, and they assisted in several gas attacks on the British front.

Field Activities.—In a sense the development of chemical warfare organisations by the Americans was deprived of its promised success. The Allies regained the general and final offensive before American plans matured. But if the latter were prevented from participating in various types of cloud and stationary attack along the front, yet the coincidence of their organisation with the development of more open warfare gave them an opportunity, which they readily seized, to demonstrate the possibilities of mobile chemical attack. Two gas companies, known as the 30th Engineers, were assembled, partially trained, and embarked for France at the end of 1917. They entered upon a course of training with the British Special Brigade R.E. while further units were being organised in America. The projector attracted the Americans, and they were ready, as General Fries informs us, to launch a big projector gas attack, when Marshal Foch's counter attack disorganised the front concerned. They then turned their attention to the use of the four-inch Stokes mortar in an attempt to neutralise the German machine-gun nests, using phosphorus for smoke and thermit shell, and continued to assist the infantry either by taking part in the preparations for attack or in subsequent operations.

Special Difficulties.—The great length of the American lines of communication led them to develop certain research and experimental organisations near to the front. These had to deal with the "short range" problems, those of immediate importance, without referring them back to America. The 3000 miles of ocean represented a necessary loss of contact which prevented the home workers, however willing, from fully realising the needs of the problems concerned. Accordingly a strong experimental station, Hanlon Field, was developed near Chaumont, and a well-equipped laboratory was established at Puteaux, near Paris.

Edgewood Arsenal.—The organisations developed in America were of very great interest. The American officers in the field, through their contact with the British and French, realised early that we were extended to the utmost in the matter of production, that our demands and programmes were far ahead of our output, and that they could not reasonably expect serious help from us, either with regard to the results or the material means of production. They, therefore, made surveys of our methods and wisely determined to concentrate on production in America. As a result, they developed the phenomenal chemical warfare arsenal of Edgewood. Had the war lasted longer, there can be no doubt that this centre of production would have represented one of the most important contributions by America to the world war. Probably had production been conceived on a smaller scale, however, its results would have materialised sooner and produced greater actual influence.

A few facts with regard to Edgewood suffice to confirm its potentialities. We learn[1] that the arsenal organisation comprised a huge chlorine plant, probably the largest in the world, various chemical plants for the manufacture of the chief chemical warfare substances adopted by the European belligerents, and shell-filling plant capable of filling a total of more than 200,000 shell and bomb daily.

Research.—Supporting this production, and in connection with the other branches of chemical warfare, a tremendous research organisation developed which, with the exception of the combined research facilities of the I.G.[2] was probably the largest research organisation ever assembled for one specific object. It grew until it contained 1200 technical men and 700 service assistants, and we are told that its work covered exhaustive research on more than 4000 different materials. Nor were the Americans less ambitious on protection. Wisely adopting the British Box Respirator during the early stages, they made vigorous attempts at the same time, with considerable success, to develop a form of their own.

Production.—An American opinion on the importance of Edgewood Arsenal at the time of the Armistice is worth quoting.[3] "Here is a mammoth plant, constructed in record time, efficiently manned,

[1] *Journal of Industrial and Engineering Chemistry*, January, 1919.
[2] The great German organic chemical combine.
[3] *Journal of Industrial and Engineering Chemistry*, January, 1919.

176

capable of an enormous output of toxic material, and just reaching its full possibilities of death-dealing at the moment when news is hourly expected of the signing of the Armistice. What a pity we did not possess this great engine of war from the day American troops first sailed for France, for, had we been so prepared, how many of our boys who 'have gone West' could have returned for the welcome home! Shall we forget this lesson of preparedness? Is this great plant to be scrapped? Possibly wise heads may find a solution of the problem which will add this great resource to American chemical industry, at the same time preserving its value to the nation as a greater asset, in case of future war, than a standing army."

Although mainly dependent on Edgewood Arsenal for their war schemes, it is perfectly clear that the Americans realised that theirs was not the ideal way, in fact was a very wasteful and inefficient way to produce poison gases or chemical warfare substances. Indeed, even during the war, in spite of their huge arsenal they established contact with various American chemical producers. At the present time, except in connection with its use for emergencies during the next few years, this huge source of production at Edgewood must be regarded as an unnecessary burden upon the State. To be of any use, it requires costly maintenance. It is only capable of producing a limited number of organic substances. Some of these are likely to become obsolete as time goes on. This reliance upon a huge fixed arsenal is not only out of accord with any international scheme for disarmament, but it is altogether too ponderous, and not sufficiently flexible for reliance in future emergencies. This is fully realised in America. General Fries, addressing the American Chemical Society, said: "The magnificent plant at Edgewood may soon be a thing of the past. We do not believe the Government should attempt to manufacture poisonous gases on a huge scale." He explains how, by reliance upon normal chemical industry, "We believe we can build up more quickly and to a greater extent than by any other method the necessary large output of poisonous gases required in a war with a first-class Power." Referring to the mobilisation of industry for this purpose, he says: "We believe that if this is done satisfactorily it will be one of the greatest possible guarantees of future peace."

Post-Armistice Developments.—But perhaps the most interesting and significant aspect of American chemical warfare development concerns what has occurred since the Armistice. Valuable and successful attempts have been made to educate not only the public but also political leaders to its real meaning. No one examining the American daily and scientific press, or reading the records of the various Government Committees on the proposed bills of chemical, or chemical warfare, interest can doubt that the Americans are probably as a whole much more alive to the importance of this matter than any other ally. Discussions on the Longworth Bill and on the new chemical warfare service have provided full ventilation for the facts of the case, in their proper setting.

It was a striking contrast to land in America early in 1920 and find New York plastered with recruiting posters setting forth the various reasons why Americans should join their Chemical Warfare Service. It was not only a sign of American methods but also one of their appreciation of the importance of the matter. This is amply borne out by their final step in reconstruction during the last few months. A separate Chemical Warfare Service has been reorgan-

178

ised in America in such a way as to give it a position of independence equivalent to that of the older branches of the service. The specific possibilities in the development of this form of warfare are acknowledged by the action of the American Congress, and this result is very largely due to the creation of an intelligently informed political and public opinion. Large grants of money have been placed at the disposal of the new Chemical Warfare Service, and its research facilities promise to equal the war establishments of the older services of other Allied countries.

Views of General Fries.—In view of the creation of this independent Chemical Warfare Service in America and of its importance when measured in terms of financial and material facilities, it is of interest to summarise some of the views already expressed by General Fries,[1] the head of the new service. With regard to the general function of chemical warfare, he tells us: "In the first place, chemical warfare is a complete science in itself. No other invention since that of gunpowder has made so profound a change in warfare as gas is making, or will make, in the future.

"To-day there are only four really distinct arms of the Service, viz.: the Infantry, the Artillery, Aviation, and Chemical Warfare. All other forms of warfare are a combination, more or less complete, of these. The gases, smoke, and incendiary materials that make up chemical warfare are used to a greater or lesser extent by other arms, but wherever gas is used it compels precautionary measures that are found in no other branch of the Service.

"Considering its power, it has no equal. Physical vigour is one of the greatest assets in any army. Gas, used properly and in quantities that will be eas-

[1] *Journal of Industrial and Engineering Chemistry,* 1920.

ily obtainable in future wars, will make the wearing of the mask a continuous affair for all troops within two to five miles of the front line, and in certain places for many miles beyond. If it never killed a man, the reduction in physical vigour, and, therefore, in efficiency of an army forced at all times to wear masks, would amount to at least 25 per cent., equivalent to disabling a quarter of a million men out of an army of a million."

The Gas Cloud Inescapable.—He goes on to explain some of the more specific military needs which can be met by chemical means, and refers independently to a point which the Germans have mentioned repeatedly in their memoirs. "One great reason why chemical warfare will continue is that it fills a long-felt want on the part of the soldier, that of shooting successfully around a stump or rock. The gas cloud is inescapable. It sweeps over and into everything in its path. No trench is too deep for it, no dug-out, unless hermetically sealed, is safe from it. Night and darkness only heighten its effect. It is the only weapon that is as effective in a fog or in the inky blackness of a moonless night as in the most brilliant sunshine. Only the mask and the training that go with it protect. Terror, confusion, lack of discipline and control are fatal."

Importance of Smoke.—General Fries is insistent on the future importance of smoke in warfare:

"Chemical warfare includes gas, smoke, and incendiary materials, and they can't well be subdivided. As before stated, all the early gas attacks were in the form of clouds. The value of that cloud, not only for carrying gas but for screening purposes, began to be realised in the fall of 1917. Clouds of smoke may or may not be poisonous, and they will or will not be poisonous, at the will of the one producing the smoke.

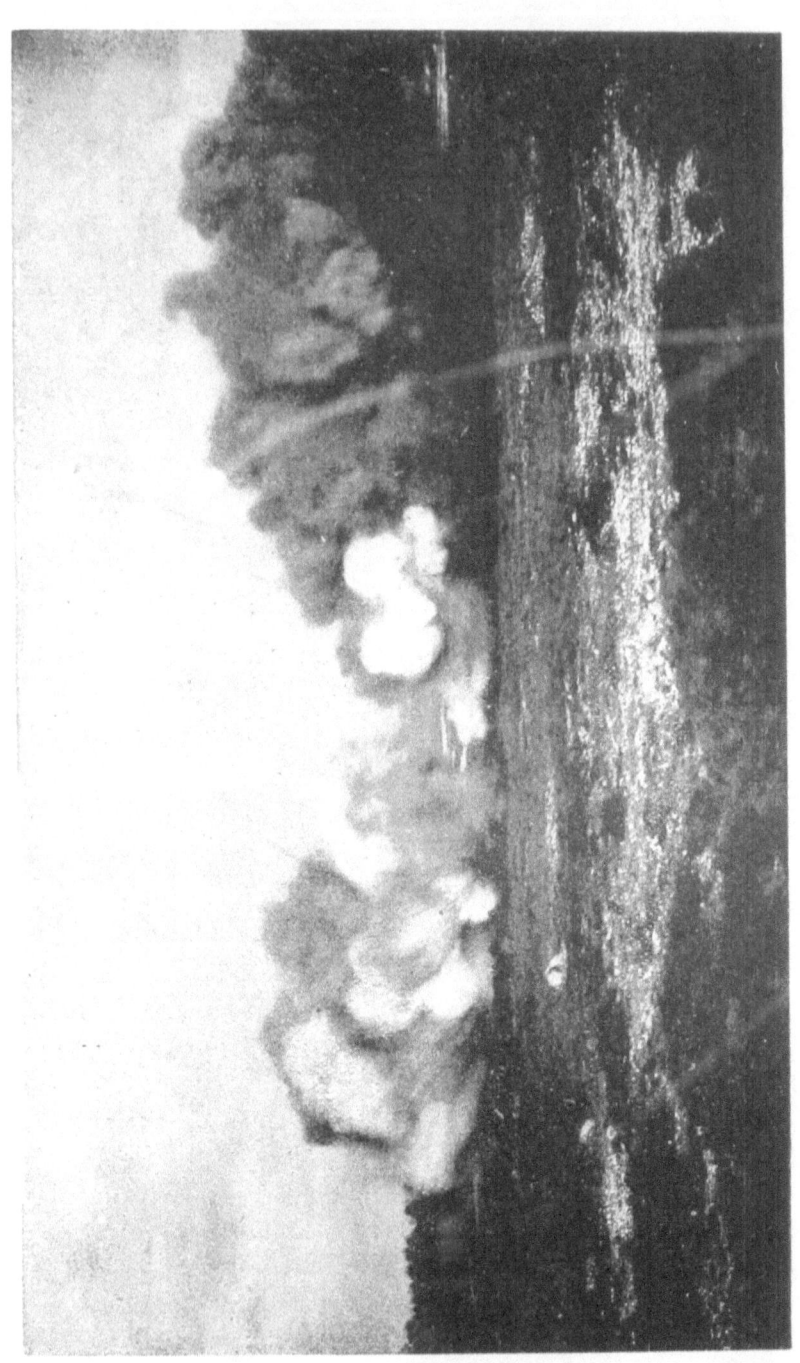

SMOKE BARRAGE.

Note the sharp curtain which is formed, behind which the infantry advance.

To face page 181

For that reason every cloud of smoke in the future must be looked upon as possibly containing some deadly form of gas. When you consider this for a moment, you can realise the tremendous possibilities for ingenuity that gas and smoke afford the attacker.

"The American, trained for 300 years in meeting nature on her great plains and in her vast forests, was early appealed to by this side of chemical warfare. As early as November 3, 1917, the United States was urged, in a cablegram from the Chemical Warfare Service in France, to push the development of a large phosphorous supply for use in smokes. Not only were the early intuitions of the value of gas borne out by later events, but to-day the future of smoke appears greater still. The battle-field of the future will be covered with smoke—not the all-pervading black smoke of the battles of the Civil War and of earlier wars before smokeless powder came into use, but a field covered with dots and patches of smoke, big and little, here and there and everywhere.

"Every man who has hunted ducks and been caught in a dense fog with ducks quacking all round, and who has tried to get ducks by firing at the quack in the fog, can realise the difficulty of hitting a man on the battlefield when you cannot see him, and have only a quack, or less, by which to locate him. The smoke will be generated in candles of two or three-pound cans that can be thrown out in front of trenches; by knapsacks that can be carried and which will give off dense white smoke in large volume for many minutes; by grenades which, while they may be thrown by hand, will generally be fired from rifles; by artillery shells reaching ten, fifteen, twenty miles back of the main battle line; and finally, from aeroplane bombs whose radius of action is limited only by the size of the earth. And thus smoke becomes one of the great elements of war in the future. It is more or less wholly protective in its nature, but as it costs more and takes longer to train a man in the various problems involved in modern war than ever before in this history of the world, it is worth while taking every precaution to protect him, once you have him trained."

Casualty Percentages.—He also brings out very clearly the unique possibility possessed by gas warfare of increasing its military efficiency, while decreasing its relative atrocity:

"The death rate in the first gas attack was probably in the neighbourhood of 35 per cent. of all casualties—and everybody in front of the wave was a casualty. With the development of masks and training in the use of the mask and in taking advantage of the ground, the death rate fell. At the same time the total number of casualties fell, but not at all in the same ratio as the decrease in the death rate. From a probable death rate of 35 per cent. in the first attack it fell to 24 per cent., then to 18 per cent., and, as gas attacks by artillery became general, to 6 per cent., and finally, with the extended use of mustard gas, the rate fell to 2.5 per cent. or less."

Again referring to casualties, he gives us the startling fact that 75,000 out of the 275,000 American casualties were caused by gas, "And yet," he says "the Germans used it in a halting, comparatively feeble manner."

Short Range Projectors.—Very much alive to the future of the short-range projectors developed in connection with gas warfare, he tells us, "The Gas Regiment in the St. Mihiel battle fired on the Côte des Esparges one hundred of these high explosive bombs at the zero hour on the morning of the attack. That hill, famous for its strength through four years of struggle between the French and Germans, dis-

182

appeared completely as an enemy standpoint. Nothing remained but torn and broken barbed wire, bits of concrete pill-boxes, and trenches filled with debris, and a few scattered fragments of clothing.

"The gas troops will, in the future, handle all short-range methods of firing gas, smoke, or high explosive. They will deliver the greatest quantities of material possible up to ranges of a mile and a half or a mile and three-quarters. So effective and so efficient are these short-range methods of projection that the No-Man's-Land of the future will be the width covered by these projectors and mortars. They can't, and never will, compete with the artillery, where range and great accuracy are the most important factors. The efficiency of artillery gas shell or artillery smoke or high explosive shell is only one-fifth that of the projector. Hence, for economy and efficiency, the artillery will be used to fire gas, smoke, high explosive, and incendiary materials only at ranges beyond those reached by the gas troops."

Again, showing how the American authorities were seized with the importance of the matter, we read:

Vast Expansion in Personnel.—"So greatly were these possibilities appreciated in the summer of 1918 that the number of gas troops authorised for use against the Germans was increased from six companies to fifty-four. Back of all this, however, was the productive capacity of the United States, which ensured that those troops would be able to fight day and night, summer, winter, and fall, until the war was over. No wonder the German quit—it was time, and he knew it."

And in conclusion General Fries tells us:

"The universal adoption of gas warfare on sea and land and in the air, combined with its persistent quality, will make that nation able to produce and use gas in the largest quantity superior in war to any other nation on the globe. The United States can reach that position and maintain it, and I believe that we are going to get such encouragement from the War Department that we can do it. I feel sure that the army appreciates the value of chemical warfare, and that it appreciates also the value of the chemists to chemical warfare.

"So long as there is any danger of other nations continuing these methods of warfare, research and experiment in chemical warfare must be pursued. Research must not only be directed towards the gases and apparatus likely to be employed in the future, but also towards protection against all possible gases. Training in the use of gas will be confined to appropriate branches, but training in defensive measures will include the whole army.

"We must continue our studies of what is known as chemical warfare. No nation has renounced the use of poison gases as the result of the Peace Conference. There are nations whose word we could not respect if they did renounce it. It is essential to study the offensive side of chemical warfare if we are to be prepared for defence. The great importance of adequate defensive appliances arises from the fact that preparations for the offensive use of gas can be made in peace-time with great secrecy, and may have far-reaching and even fatal results in the early stages of a war.

". . . For these reasons it is necessary to make adequate provision for research, experiment, and design in connection with war material. It is equally necessary to avoid overlap, duplication of effort, and the setting up of military institutions for scientific research which can better be done by existing civil institutions."

184

He also quotes from a statement from General Debeney, Director of the French College of Warfare:

"Should war begin now, aviation, and especially gas, would play one of the most important parts. The progress of aviation would make the rear of each front, and very far in rear, extremely dangerous, and the progress of chemistry would permit the use of gas on zones of such an extent as cannot be imagined,

"Making gas is naturally rapidly done, because all the manufacturers of chemical product—still so numerous in Germany—can be requisitioned, but to make airplanes is much slower.

"The defence against gas seems to be more difficult than against airplanes. I believe that against airplanes, the anti-aircraft artillery is susceptible of making rapid progress, and perhaps in that very instance gas will be one of the best ways, if with appropriate shells *the air can be poisoned all around the attacking airplanes.*

"It would be much more effective to create, for example, a sphere of poisoned air a mile round the airplane, instead of trying to hit the machine directly with bits of the shell."

British, French, and even German opinion, while not underestimating the importance of the matter, may not agree in an unqualified way with all the above statements. But we claim that they show vision in a branch of war which, on account of its scientific basis, may, more than any other, speedily prove the visionary a true prophet.

CHAPTER IX

GERMAN CHEMICAL POLICY

The preceding account of chemical warfare leaves the impression of a successful Allied struggle against persistently unfavourable circumstances. We were constantly compelled to accelerate to attain the pace set by the enemy. There were exceptions, undoubtedly, but in the main Germany kept ahead in the chemical struggle.

So far, in examining the root of our troubles, we have been content to refer to the existence of the I.G., to describe its chemical warfare activities, and to indicate, briefly, its unique power to produce large quantities of organic chemical products at short notice. The close connection between the German dye industry and chemical warfare is now well recognised in official circles, and, to some extent, by the general public. Its belated exposure was almost entirely due to the facts revealed by the Inter-Allied Mission to the German chemical factories some months after the Armistice.

But the situation thus revealed was not created in a day, nor by chance. Indeed, one of the military features of industrial chemical development in the I.G. has already been traced to pre-war activities. I refer to the Haber process for the production of synthetic ammonia. It would be short-sighted policy to accept the set of conditions against which we struggled, and to explain them in terms of the I.G., without looking more closely into the pre-war activities

186

of this organisation. Such an examination may reveal the basic forces which determined our inferior position in chemical warfare at the outbreak of war. It is true that we can explain away our inferiority by referring to the German breach of faith, which automatically created conditions for which we were unprepared. This is a comfortable solution. But had chemical warfare been a strongly developed and accepted method of war before the outbreak of hostilities, would we then have been prepared? The records of the past, before April, 1915, must be consulted to answer this question. We may find that our position is due to more than a mere negative attitude, to more than our simple neglect of the organic chemical industry. It may be that there were forces which definitely exploited this national characteristic to our disadvantage. The pre-war policy and activities of the I.G. must be examined from this point of view. In no country has such an investigation been more complete than in America, and official statements have been issued by the American Alien Property Custodian[1] which throw a flood of light on the pre-war activities of the constituent branches of the I.G. They conclusively reveal the existence of a carefully directed German chemical policy making for world domination in the organic chemical industry, which greatly hampered the military effectiveness of other countries, and directly strengthened the military resources of Germany. On broad lines, the pre-war and war activities of the I.G. produced the same result as an attempt to strangle the economic life of possible opponents, enfeebling their resistance to the subsequent delivery of a hammer blow designed to take maximum advantage of the

[1] *Alien Property Custodian Report*, Washington. Government Printing Office, 1919.

situation thus created. Twenty years or more under the régime of a forceful economic policy, not without its sinister aspects, prepared the ground by weakening us in the concentrated chemical warfare which ensued. The success of this policy manœuvred us into such a position that we barely escaped defeat under the hammer blows of German chemical aggression. This, in fact, appears to have been the German conception of modern war in its relation to industry, and American reports have shown that it was carried through with typical thoroughness by familiar German methods.

Origin of German Chemical Monopolies.—The completeness of our organic chemical deficiencies, and the thorough way in which we had failed to develop organic chemical industries, creates such a sharp impression, when thrown into relief by the outbreak of war, that we are led to inquire into the methods by which these monopolies were established. Let us admit, without any further delay, that Germany owed the origin and assertion of these monopolies in part to her scientific development, fostered by a vigorous policy of applying scientific research to industrial enterprise. So far as her success depended upon such factors, it merits our unqualified admiration and envy. But stimulating these developments was a very definite general and commercial policy which requires close examination.

German Chemical Commercial Policy;—Evidence of the U.S.A. Alien Property Custodian.— Giving every credit to German initiative and thoroughness in the application of science to industry, we are still prompted to inquire how this monopoly came to be so complete. We can rely on more than mere rumour, when examining the commercial methods of the great I.G. The American Alien

188

Property Custodian, Mr. Mitchell Palmer, and, later, Mr. Francis P. Garvan, had occasion and opportunity to make minute examination of the German dye agencies in America in connection with general investigations on the reorganisation of alien property. Their revelations truly merit the term, showing remarkably clearly the unity of conception, determination of purpose, and co-operation with the German Government which characterised the policy of the I.G.

Pre-war American Situation.—Let us briefly consider the relevant aspects of the pre-war American situation. According to fairly well-known facts, confirmed by the reports of the two American officials mentioned above, the American pre-war organic chemical industry consisted of little more than a series of small assembling plants. Although enormous supplies of coal-tar products were available, yet the dye intermediates derived from them were not made in America, but imported from Germany. After various attempts to establish the dye industry, it seeemd, at one time, about 1880, to have definitely taken root, but, within the space of five years, there were only four dye producing establishments remaining.

German Price-cutting; — Salicylic Acid. — In every instance the manufacture was almost immediately brought to an end by German price-cutting. The same source reveals the direct and indirect methods used by Germany to prevent, at all costs, the development of an independent organic chemical industry. There are many pointed examples of the direct method, and we will glance at the case of salicylic acid. This is a very important chemical, used not only for certain important drugs but also as in intermediate for dyes and photographic chemicals. In 1903 the United States possessed five manufacturers of this product. In ten years' time three of these had failed, and one of the survivors was a mere branch of a German house. During this fatal ten years, the product was being sold in that country at a price twenty-five per cent. lower than in Germany. The manipulation of the prices of the other products of the German monopoly enabled them, by such methods, to maintain it. Many other examples, including such important products as bromine, oxalic acid, and aniline, could be quoted to show the results of the German price-cutting policy. The direct significance of bromine for chemical warfare must be borne in mind.

Full Line Forcing.—Besides directly attacking the production of raw materials and intermediates, the Germans used an indirect method which has been described as "full line forcing." They were the sole producers of certain specialties, such as alizarine colours, anthracene colours, and synthetic indigo. These were indispensable to the textile manufacturers, and by refusing to supply them, except to houses which would buy their other supplies from German manufacturers, the latter could squeeze out home producers of simple dyes, however efficient their production.

Bribery and Corruption;—German Patent Policy.—The dyeing industry was peculiarly susceptible to corruption. It was so simple for the head dyer of a mill to show a partiality for dyes from any particular source of supply. The American Alien Property Custodian very frankly tells us[1]: "The methods of the great German houses in carrying on their business in this country were from the first honeycombed with corruption. Bribery of dyers was

[1] *Alien Property Custodian Report*, 1919, p. 34.

carried on almost universally on a large scale. . . . So extensive was this corruption that I came across only one American consumer that had escaped its ill effects." Such were hardly the methods of decent commercial competition, although it appears that the strong patriotic sense of the German was able to justify, in his own eyes, what might be regarded as reprehensible methods. This is not a question of bringing up old reproaches, but merely of coldly examining facts. We have already referred to their patent policy, whereby thousands of patents were taken out, the only value of many of them being to cramp the productive initiative of possible rivals. Professor Stieglitz explains how the German patents were useless in developing large scale manufacture. "The patent protects the product, but does not reveal the method." Sir William Pope has also brought out this point, showing how the Germans use thousands of bogus patents to protect their chemical industry. He tells us,[1] "In fact, some German patents are drawn up for the purpose of discouraging investigation by more practical methods; thus, any one who attempted to repeat the method for manufacturing a dyestuff protected by Salzman & Kruger in the German patent No. 12,096 would be pretty certain to kill himself during the operation."

Propaganda and Information;—Espionage;—Activities of the Dye Agencies.—But another method which was used in this commercial offensive, to which we must draw further attention, dealt with propaganda and information. In his comprehensive report, the American Alien Property Custodian examines a number of large industries and reveals how the German interest in these industries through their

[1] *Science and the Nation.* A. C. Seward, F.R.S. Cambridge University Press, 1917.

American ramifications were active, "sowing the seeds of German propaganda," and collecting information, both commercial and military, for the use of the German Government and its agents. Quoting again from this report, "In many of the large German-owned companies taken over by the Alien Property Custodian, after investigation it was found that espionage was one of the chief functions. Every scrap of information of commercial or military value to Germany was carefully gathered by the representatives of these concerns in this country and quickly forwarded to the home office in Germany. The German agents were particularly keen on gathering information that would be helpful to Germany's commercial warfare. Once in Germany, this information was carefully card-indexed for the use of the manufacturers. Bulletins of commercial information were also prepared and placed at their disposal. In Germany, the collection of all commercial information is under a bureau which is controlled and financed by the great German banks, such as the Dresdner, Disconto, and Reichs Bank." This statement is not mere generalisation, but is backed by innumerable examples. Thus we find a light railway equipment manufacturer, a projectile company, a wireless company, various magneto companies, insurance companies, and German shipping companies, all engaged in spreading propaganda, acquiring information, and influencing public opinion in favour of Germany. But, undoubtedly more important than any of these, and taking a leading part in the general scheme, was the German dye organisation. The American publications make this quite clear. Mr. Garvan goes so far as to say: "As long as you were supplied by the big six (*i.e.* the I.G.), your business had no secret unknown to Berlin. In Berlin you will find the card index system which recites every fact connected with each and every one of your sources which can be of any possible value to your rivals over there." Referring to assistance rendered by various American and Allied departments, including Military, Naval, and War Trade Intelligence, we learn from the same sources: "All these bodies worked in close co-operation and their mutual assistance was of inestimable value. Information derived from these sources demonstrated that the chemical industry was a natural centre for espionage and that this had been true long before we entered the war—indeed, before the war began. The relation between the German Government and the great German chemical houses was so close that representatives of the industry were naturally almost direct representatives of the Government, and their work in this country gave them unequalled opportunities for examining our industries from within."

With the outbreak of war, this organisation became more clearly defined. It was, perhaps, difficult before the war to know where to draw the line between purely commercial and actual governmental German activities. The outbreak of war left no room for doubt. The German dye agencies became, at once, the active agents of their Government in various schemes, the nature of which we shall outline, and their "information" functions became very definitely describable as espionage.

Manœuvring Raw Materials.—In the first place, the Alien Property Custodian found unexampled evidence of a definite German scheme to corner and divert certain important war materials destined for the Allies.

Chemical Exchange Association;—Doctor Albert's Letter.—Many such plots could be quoted,

193

but we will limit ourselves to one,[1] chosen because on its stage move the chief figures of this espionage system. This case has been described under the name of the "Chemical Exchange Association," and is much more fitted for the pen of a Conan Doyle. The move appears to have been initiated by Dr. Albert, the financial adviser of the German Government in America, in collaboration with von Bernstorff. Its purpose was to corner the immediate supplies of American phenol in order to prevent its manufacture into high explosives, including the well-known picric acid. The outbreak of war instantly stopped the entry of phenol into the country. Further, this product was not manufactured there to any extent before. Large supplies were required for the production of synthetic resins, for the gramophone industry. This led to the development of a phenol industry by the Edison works, and there appeared, automatically, a phenol surplus. Dr. Albert, aware of the probable fate of this surplus as raw material for allied munitions, determined to seize it for the German Government, and he did this through Dr. Hugo Schweitzer, one of the most prominent members of the American agency of the great Bayer works. In June, 1915, Dr. Schweitzer contracted with the selling agents of the Edison Co. for the entire surplus of phenol available for sale, offering a large cash security which was furnished by Dr. Albert. A lapse of a week witnessed another contract with the Heyden Chemical Works, a branch of the German house, by which this phenol was purchased for conversion into salicylic acid and other products. To avoid exposing the nature of the deal, Dr. Schweitzer registered as the "Chemical Exchange Association." The profits amounted to

[1] *Alien Property Custodian Report*, 1919, p. 43.

nearly a million dollars, half of which belonged to Dr. Schweitzer. This, we are told, went immediately to the German Government. As a suitable climax to such a venture, a dinner was given at the Hotel Astor by Dr. Schweitzer in honour of Dr. Albert, and is described as a typical gathering of the most active German propagandists in the country. It was as a result of this deal that Dr. Albert sent Dr. Schweitzer a memorable letter in which he praises his "breadth of highmindedness," and compares his work with "a military coup accomplished by an army leader in destroying three railroad trains of forty cars containing four and a half million pounds of explosives."

Dye Agency Information System;—Dr. Albert on Chemical Warfare.—Although a great deal has been said in America with regard to the activities of Dr. Schweitzer and his followers, very little has been heard on this side. Explaining the complete information system possessed by the Germans, Mr. F. P. Garvan informs us that the head of the system in America for years before the war was Dr. Hugo Schweitzer, President of the Bayer Company there, and he even quotes his secret service number given him by the Imperial Minister of War, stating that he came to America, became a citizen on the instruction of the German Government, and led the espionage and propagandist movements down to the day of his sudden death in November, 1917. The relationships between Dr. Albert and Dr. Schweitzer, when the former was leaving for Germany in 1917, are very illuminating. We learn from the same source how Dr. Schweitzer received from the former nearly one and a half million dollars, all to be spent in espionage and propaganda. Dr. Albert, leaving Dr. Schweitzer a letter of appreciation, to which we have referred in connection with the Chemical Exchange, makes a very significant reference to chemical warfare. "Of still greater and more beneficial effect is the support which you have afforded to the purchase of bromine. We have a well-founded hope that, with the exclusion of perhaps small quantities, we shall be in a position to buy up the total production of the country. Bromine, together with chloral, is used in making nitric gases, which are of such great importance in trench warfare. Without bromine these nitric gases are of slight effect: in connection with bromine they are of terrible effect. Bromine is produced only in the United States and Germany. While, therefore, the material is on hand in satisfactory quantities for the Germans, the Allies are entirely dependent upon importation from America." Making due allowance for the fact that Dr. Albert was not a technical man, this information possesses an element of truth, indeed France was driven to the extreme of establishing a bromine industry in the wilds of Tunis in order to counter the German attack.

The Moral Aspect.—Such facts tempt us to think hardly of these representatives of German culture. But they were, no doubt, fiercely patriotic Germans, and it is not difficult for us to understand their activities after the outbreak of war. An American, however, can hardly adopt such a lenient view, if, as has been claimed, many of these agents were naturalised Americans, for they were abusing the privileges and the confidence of their adopted country. We have no wish, however, to dwell on this aspect of the matter, and have no doubt whatever that many good Germans could justify all these activities according to their own codes. It would have been

196

better not to have given this information the light of day, were it not of some value for the future.

Report of the New York World;—German Policy Regarding Dye Supplies to the U.S.A.— How far can the parent organisation of these dye agencies be regarded as aware of their activities? They were largely responsible for their inspiration. Mr. Garvan says, "Practically all the dye salesmen were only nominally in the employ of the branches here; all had secret and personal contracts with the Home Office." From these facts alone there can hardly be any doubt as to the connivance of the home organisation. Again, on April 28, 1915, the *New York World* printed an editorial explaining that "two large German chemical and aniline dye concerns are reported to be establishing factories in New Jersey, to supply American demands hitherto supplied from Germany." This statement apparently alarmed Captain Boy-Ed, the German Naval Attaché, and he communicated with Dr. Albert, the financial representative in New York, for the establishment of these factories would have countered the German policy of bringing political pressure by refusing dye shipments. Dr. Albert's reply to Boy-Ed contains the following phrase: "With regard to the dyes, I got into touch with local experts in order to determine what truth there is in the news. According to my knowledge of things, the matter is a fake, inasmuch as *our factories have bound themselves orally and by word of honour to do nothing in the present situation which might help the United States.*" As further evidence of this definite policy, witness a letter from Consul-General Hossenfelder to the Imperial German Chancellor, Dr. von Bethmann-Hollweg. This letter is dated New York, March 3, 1916, and, after a detailed examination of the economic relationships between Germany and America, states: "Further, we should, according to my conviction, hold ourselves absolutely passive in relation to the proposals for the exportation of potash, chemicals, and dyestuffs, and if the opportunity arises, make the sanction for them, not dependent upon the consent for an exchange of articles, but upon the abolition *en bloc* of all hindrances to intercourse contrary to international laws which have been instituted by England." Further, Dr. Albert, cabling to the German Government in April, 1916, on the export of dyestuffs, tells us: "The hope was entertained of bringing American industries which were solely dependent upon German deliveries of dyestuffs into a position that they would have to insist on the importation of dyestuffs under the conditions demanded by Germany." There can then be no doubt that the parent organisation of the I.G. was in close touch with the activities of its agencies.

This, then, is a brief account of the methods by which Germany created the monopoly whose existence threatened our success in the world war. Before leaving the question of the monopoly, let us inquire a little more closely into its exact nature and range. Various American official reports have revealed the desperate measures necessitated in that country in order to meet deficiencies in vital products when the German source of supply was removed.

Professor Stieglitz's Evidence.—Professor Stieglitz, of the University of Chicago, giving evidence before the United States Senate, stated:[1]

"I have come to the conclusion that we would have saved a great deal of suffering and a great many lives in this country, if we had had an organic chemical industry, as they have in Germany, before we started

[1] Hearings before the Committee on Finance, U. S. Senate, 1920.

the war." Characterising the dye industry as the source of war chemicals, including explosives and poison gas, he emphasises the drug question and shows how their development depends absolutely upon the existence of certain raw materials, and facilities for comprehensive organic chemical research, which only find a *raison d'être* in the existence of a flourishing dye industry.

Ehrlich's Discovery.—Pointing out the difficulties in developing the manufacture of salvarsan, he explains how the process was originally discovered by an organic chemist, Dr. Paul Ehrlich, co-operating with a German dye company, the crude material coming from the dye plants, the product itself strongly resembling dyes, "containing arsenic instead of part of their nitrogen." The great importance of this drug is brought out by another witness before the same committee, Mr. Francis P. Garvan, who explains how, by refusing or neglecting to ship salvarsan, Germany wanted the United States "to starve to death" for lack of it, and he continues: "Think what an extension of disease and what an intensification of suffering and distress Germany was willing to impose upon her best market in order to obtain her imperial will."

Germany had monopolised the production of the important synthetic drugs, including the derivatives of salicylic acid, of which aspirin had developed wide use in Allied countries. After every household had learnt the value of German produced aspirin, its supply was cut off at the outbreak of war. The same disadvantages applied in the field of anæsthetics. For a long period America had no local anæsthetics for hospital surgical work, being compelled to use what were termed "Bulgarian Operations," that is, operations without anæsthetics. Pro-

199

fessor Stieglitz claims that the lack of drugs and anæsthetics threw back American surgery some fifty to seventy years in civilisation.

But what of this country? We have already outlined how the outbreak of war found us with, at the most, two or three relatively small producing centres, which did valiant service during the war and amply proved the importance of the dye industry by revealing what could have been done had we been many times stronger. Was the same German chemical policy responsible for our pre-war position? As far as we know official investigations have not been pursued to the same length as in America, but it is beyond doubt that the German dye companies took every possible step to stifle the development of our organic chemical production. When the war broke out, our comfortable commercial contact with the I.G. became a strangle-hold. It could not be otherwise. Whatever the German attitude, and we could hardly expect it to be friendly, the stranglehold at the outbreak of war was inevitable. But this dye menace facing our textile industries, and weakening our power of retaliation in the chemical war, was not the only danger from the I.G. We were in a critical position through failure to produce other commodities than dyes.

Drugs and Medicinal Products;—The German Monopoly;—National Health Insurance Commission.—The question of drugs assumed critical importance at the outbreak of war. Germany had been asserting her monopoly for years in the field of medicinal chemicals. Cessation of supplies at the outbreak of war caused grave apprehension of a serious shortage in these products, so important for the adequate treatment of disease. In some cases we possessed neither the raw materials nor the technical knowledge to undertake rapid home production. But in the important group of the synthetic drugs derived from coal-tar products, the raw materials were produced in quantity in the United Kingdom, only to be exported to Germany, thus contributing to her monopoly. British manufacturers, on the other hand, held their own in the production of certain kinds of drugs, such as the alkaloids, gaseous anæsthetics, and some inorganic salts of bismuth and mercury. In a summary of certain war activities of the National Health Insurance Commission, we read: "It was chiefly in the making of the coal-tar synthetic remedies that Germany was pre-eminent, and that position was due not to any lack of skill or invention on the part of the British chemists, but to the high degree of organisation attained by the German chemical industry, which made it possible to convert the by-products of the aniline factories into medicaments of high therapeutic and commercial value."

The Royal Society;—Novocain.—So serious was the situation that for some time we existed on feeble stocks. But during this period the utmost efforts were made to develop our own production. The Royal Society promptly came forward with a scheme to link up would-be producers with appropriate centres of research. The latter not only assisted production but actually produced sufficient quantities of important drugs to tide us over the difficult period. Thus, for example, for the production of novocain the assistance of about forty university laboratories throughout the country was invoked, and they proceeded to produce the intermediates, diethylamine and ethylene-monochlor-hydrin. These substances were converted into diethyl-amino-ethanol, and the final step, the production of novocain, was undertaken

The Riddle of the Rhine

by manufacturers, including a prominent dye firm. We have referred to one of these substances in connection with the German production of mustard gas, and need only say that in England, in a time of national emergency, the Government had to depend on the improvised assistance of forty teaching and research institutions for the production of small quantities of drug intermediates. Further, this work, although to the permanent credit of those who undertook it, did not enable us later to produce rapidly war quantities of mustard gas, itself dependent on the same important intermediate, ethylene-mono-chlor-hydrin. Germany settled the drug and mustard gas question by a simple demand to the I.G., because the latter, holding the indigo monopoly, possessed actual large-scale ethylene-chlor-hydrin production.

Other cases, although equally creditable to those actually engaged in the work, also reflect our national unpreparedness and neglect of chemical industry.

Beta-Eucaine.—Beta-eucaine is a very important local anæsthetic. Before the war we obtained it almost exclusively from Germany. When urgently needed in 1915 for the War Office and Admiralty, the Government discovered that it could not obtain this substance from commercial sources. Seventeen laboratories co-operated to produce two hundred and sixteen pounds of the material. Such examples would be ludicrous did they not possess such a serious national aspect. Our position was almost as desperate regarding chloral-hydrate, the important hypnotic, and the rare carbo-hydrates required for bacteriological purposes. Sir William Pope's comprehensive statement[1] supplies further examples.

Photographic Chemicals.—Our dependence upon

[1] *Science and the Nation*, A. C. Seward, F.R.S. Cambridge University Press, 1917.

German monopoly, so drastically revealed at the outbreak of war, was not limited to dyes and drugs. Photographic chemicals were of special importance for war purposes, yet, when the development of aviation increased our demands for photographic chemicals, we had no normal sufficient source to which to turn. We needed not only the essential bulk chemicals, such as amidol, metol, para-amidophenol, and glycine, but also certain rarer substances, such as the photographic sensitisers, which were so essential for the Air Force. By calling upon chemical industry and research institutions both needs were satisfactorily met, but the contrast with Germany leads perforce to the same conclusion, their ease and speed of production as compared with ours.

This examination shows the fine texture of the tenacious web by which Germany had entangled and stifled the organic chemical industries of other countries. Although at the outbreak of war the Allies were slow to realise the war significance of the dye industry, yet they were quick to determine that the resumption of peace would not find them in such an ignominious position. Steps were taken to establish dye industries in England, France, and America. Not only did plants spring up to meet the immediate needs of the textile industries of the world outside Germany, but the question received considerable Government attention. Promises were made and steps taken to encourage the growing industries. But these cannot be examined in detail here, and the main facts are common knowledge. Two points emerge, however, which are of prime importance from the point of view of our discussion. In the first place, the acute needs of the armies prevented the maximum use of the war opportunity for developing Allied dye industries on a sound basis. No sooner was producing capacity installed, than it was taken over for the production of urgently needed organic chemicals for explosives. Dye enthusiasts would have regarded the war as a supreme opportunity for a period of concentrated organic chemical research to make up the leeway which existed, owing to forty years of German development. But the research energies of the country were occupied on more pressing problems. In Germany, the war chemical activities of the dye factories all contributed to their future post-war strength. In England and France it was otherwise. Our equivalent energies were concentrated on developing improvised processes and plant, absolutely necessary to counter the German attacks, but almost without exception of no direct ultimate value to our peace organic chemical industries. This is a point which merits careful consideration. These industries voluntarily threw aside what was, logically, a great opportunity for them to push their research investigations so necessary for eventual success. The state-aided Huddersfield factory represented national vision, whose fruits were stolen by our ceaseless need to improvise counters to German aggression. But we owe to our dye industry the national recognition of these facts. Stress of war gave us true vision, but prevented its logical outcome. War needs are now removed, and everything should be done to place at the disposal of the dye industries those facilities which they necessarily, but gladly, sacrificed in time of emergency.

The brief survey of the preceding pages reveals the existence of a German chemical policy pursued vigorously for many years before the war. It also shows how this policy developed in America, the chief neutral country, during the war period, for two years before her entry.

The Americans have also established beyond doubt the active co-operation between the German Government and the I.G. But, if the policy of the German Government and of the organic chemical industry had many points in common before the war, they became one before hostilities were many months old. The part played by the I.G. in munitions production, in which it was virtually a tool of the Government, has already been seen. It must be remembered that, after the first Battle of the Marne, the German Government turned to the I.G. for a large part of its explosives and practically all its poison gas, and, as has been stated on many occasions, and with reason, Germany would not have been able to continue the war after the summer of 1915 but for the commercial development of the Haber process by the I.G. The story is too well known to repeat at length. The basic element of explosives is nitrogen, which is introduced by nitric acid. This was produced from imported Chili saltpetre, but the blockade cut short these imports, and but for the Haber method, the vital step in producing nitric acid from the air, Germany would have been compelled to abandon the struggle.

There is striking coincidence between the commencement of the Great War and the successful completion of certain vital German chemical developments. As late as 1912 Germany still depended on other countries, chiefly England, for her phenol, the basic raw material for picric acid as well as a dye necessity. Soon after that date the development of the Bayer plant made her independent in that product, and gave her, in fact, an exportable surplus.

War Activities of the I.G.—Reviewing all these activities and realising how they all emanate from this one organisation, we are overwhelmed by its formidable nature as an offensive and defensive weapon in time of war. Here we have an organisation, the I.G., whose sinister pre-war ramifications dominated the world by their hold on the supply of organic chemicals vital for peace and war. This organisation functioned, in a sense, as the life blood of German offensive warfare. German sources tell us very little of the war activities and future significance of the I.G. A veil of secrecy seems to be cast over the whole matter, but behind this veil must exist an acute realisation of the value of the I.G. as a trump card for the future. Krupp is uncovered, the whole world was alarmed at its meaning for war, but heard with a comfortable sense of security how Krupp was exchanging the sword for the plough. But the gigantic I.G. controls in its great hand a sword or plough for war or peace at will. This is no far-fetched metaphor.

The Rhine Factories and the Armistice.—It therefore becomes important to inquire into the attitude and activities of the I.G. since the Armistice, and to examine its position in world reconstruction. For one brief period, the few weeks following the Armistice, the German dye industries appear to have been without policy, its leaders in confusion. But with the confidence inspired by the Allied Rhineland occupation, with the assistance provided by the Allied controlling organisations with regard to labour, fuel, and commercial transactions, the industrial morale speedily recovered.

The tide of revolution which accompanied the German debacle in the autumn of 1918 swept over the Rhineland chemical factories. Colonel Norris, writing on his visit in February, 1919, tells us that

after peace was restored by the Allied forces:[1]—
"the managers of several factories agreed that the
occupation of the territory was the best thing that
could have happened. On the other side of the
Rhine, labour refused to work, and demanded un-
heard-of pay—everything was topsy-turvy. In fact,
before the Allied armies arrived, revolutionary no-
tions were growing rapidly along the Rhine. One
director of a well-known chemical plant is said to
have escaped by night with his life by way of the
river, when his employees were especially menacing.
When the British Army came he returned, and is
now at his old post." Thus, although the I.G. was
model in its institutions for the welfare of employees,
at least one of its most prominent directors was
compelled to take refuge from infuriated labour.
What with danger from the latter, and the uncer-
tainty of action by the oncoming Allied troops, the
future of the factories appeared very gloomy. In
fact, there are fairly credible rumours that the German directors were willing to dispose of their assets
to the Allies while they remained intact. But the
same Allied troops, whose advent was feared, rolled
back the tide of revolution from the banks of the
Rhine, and restored industrial security. It is doubtful whether the investing armies realised the full war
significance of these factories, except the French.
The latter instituted a fairly thorough control almost at once. But, judging from reports of different missions to these factories, we were even back-
ward in organising inspection of the purely munitions
plants. Thus the Hartley Mission did not materialise until three months had elapsed.

War Mentality of the I.G.—We watch a vivid

[1] *Journal of Industrial and Engineering Chemistry*, Vol. XI.,
1919, page 817.

impression of the war mentality of the I.G. in a few phrases from Colonel Norris's account: "Around the walls of the director's room was a beautifully painted and artistic frieze which pictured the various plants of the Bayer Company and their activities. Dr. Duisberg, the director, pointed out proudly to the Americans the view of the company's plant on the Hudson River. We were not surprised to see it, although pre-war advertisements had assured us at home that Bayer aspirin had been made on the Hudson for years by an American company. During the war an ante-room had been decorated in a similar way, with pictures illustrating the activity of the plant in the preparation of war-gas materials. One saw how gas was made, shells were filled, and gas masks assembled. The work was done by an artist, and has a permanent value. The fact that the thing was conceived and executed during the stress of war throws an interesting sidelight on German character." Incidentally, it also throws a further sidelight upon the part played by Leverkusen in the chemical warfare campaign.

German Attitude towards Inspection.—As was quite to be expected, the German factories did not receive our missions with open arms, and they were particularly jealous of any inspection at Oppau, the site of the wonderful Haber synthetic ammonia plant. Lieut. McConnel, of the U.S. Navy, tells us:[1] "Upon arrival at the plant the Germans displayed a polite but sullen attitude. They seemed willing to afford the opportunity of a cursory inspection, but strongly objected to a detailed examination. On the third day of the visit the writer was informed that his presence had become a source of serious

[1] *Journal of Industrial and Engineering Chemistry*, Vol. XI., 1919, page 837.

objection and that if his examination were prolonged a formal complaint would be submitted to the Peace Conference." The Allies had only themselves to blame. Their facile yielding to the argument that this great arsenal was principally of peace significance, owing to the fertilisers which it would eventually make, and the feeble backing provided for inspecting missions, were reflected in the semi-resistant attitude of the I.G. personnel.

The Rhine and Chaulny Contrast.—It was a curious contrast, however, to pass through Chaulny on the way to the Rhine. At Chaulny, the oldest chemical works in France, quoting again from Colonel Norris, "where Gay-Lussac did his famous work on the manufacture of sulphuric acid, where Courtois discovered iodine, and where plate glass was first made, had grown with the times, and was amongst the largest factories in France. Around it was a thriving town of about 13,000 inhabitants, with some excellent public modern buildings. When the Germans in their first retreat were forced to leave the place, they dismantled the factory and carried away everything that was portable. The fortunes of war brought them back, and before they left a second time a regiment of soldiers was put to work to destroy systematically the factory and the entire town. For a month they kept at work, and when they withdrew but a few bricks were left standing. Every boiler had been blown up with dynamite, and every tank too heavy to be carted away rendered useless. About half an acre was covered with chemical stoneware of all kinds; each piece had been broken with a sledgehammer. Nothing was too small or too large to escape destruction. And to make sure of a good job, everything that would burn was set on fire." Yet within twenty-four hours one met Germans, in-

directly or directly responsible for this policy of
destruction, resenting peaceful Allied inquiries on the
munition activities of their own plants. We hardly
know whether to attribute such effects of Allied
policy to our own integrity in respecting the peace
activities of these arsenals or to official ignorance of
their war-like nature.

German Revolution and the Industrial Leaders.
—It is curious how the leadership of the captains of
German industry was left untouched by the revolutionary disturbances of the post-Armistice period.
Evidence is to be found in the composition of the
main German delegation to Paris for the settlement
of the Versailles Treaty. Many of the members
were big industrial magnates, several had direct connection with chemical industry, and at least one was
a prominent director of the I.G.

The German Peace Delegation.—Commenting on
the composition of the main German delegation in
the spring of 1919, we find the German press deplor-
ing the omission of any "visible representative" of
Army or Navy. Does this imply the presence of invisible representation? Whether intended or not,
there is truth in the implication. The list contains the
name of one of the leading representatives of the big
dye combine. Others of the delegates have chemical
interests. This is significant. It more than implies
the German official acknowledgment of the importance of the dye industry in general for the future of
Germany, and of its prime importance for war.

Recent Signs of Government Interest.—Recent
developments have merely strengthened the dye combine and provided further evidence of Government
interest in its welfare. The chief signs of reviving
German Government interest in the I.G. are to be
found in the loan for the nitrogen enterprise and in

the privileges which it enjoys with regard to Government taxes. An American source,[1] a witness before a Senate Committee, reveals that the dye plants "have to pay no direct Government taxes. According to an understanding with the present Government, all organic chemical productions, the companies themselves, as well as all dependencies, without exception, for the next ten years, are freed from all direct State tax. In so far as community taxes come into consideration, I believe we will obtain a remission for our profession." The latest sign of Government support is to be found in the preferential treatment obtained by the German dye industry in coal deliveries. Coal is a critical factor in the German attempt to regain their monopoly.

Nitrogen Fixation.—The industrial fixation of nitrogen by Germany to form ammonia has great importance from the point of view of our discussion. Statements by various prominent Germans, such as Dr. Max Sering, of the University of Berlin, and Dr. Hugo Schweitzer, already referred to, leave no doubt. The former, writing in 1915, tells us: "The complete cutting off of the supply of Chili saltpetre during the war has been made good by our now taking nitrogen directly out of the air in large factories built during and before the war. With extraordinary rapidity the question has been solved how the enormous quantities of the needed ammunition were to be produced, a question which in England still meets with difficulties, in spite of the help from America."

The German Nitrogen Syndicate.—The two great Haber plants at Oppau and Merseburg are both constituent parts of the I.G., and they introduce a new element of Government interest into the I.G.

[1] Hearings before Committee on Finance, U. S. Senate, 1920, page 195.

policy. Giving evidence before the Committee on Agriculture and Forestry of the United States Senate, Colonel Joyce develops this question of Government interest in detail. He tells us how war nitrogen supply was energetically and specifically fostered by the German Government through an Imperial Commissioner under the War Department. One of the three advisers of this compaign was Doctor Bueb, representing the Badische Anilin- und Soda-Fabrik. Colonel Joyce tells us: "That was a strictly war control organisation, but even before the war closed, Germany, with her usual foresight, was giving consideration to the future commercial aspects of her nitrogen works, and in August, 1919, there was definitely formed an association of the producers which was called the Stickstoff Syndikat G.m.b.H. or Nitrogen Syndicate. This designation is a commercial one, and the organisation is along commercial lines, but it is reliably stated that the establishment of this syndicate was largely due to governmental influence. This will be more easily understood if it be realised that the German Government had given financial assistance to many of the new plants and plant increases which the war had necessitated."

Haber Process Prominent.—The Badische Co. holds a large part of the capital stock of this syndicate, whose Board contains a Government nominee. In addition the Board of Managers will have a Government chairman. Through such arrangements, Government interest in the I.G. nitrogen enterprise is clearly revealed. In conclusion, Colonel Joyce informs us, "This information, which comes from most reliable sources and is not to be disputed, shows that, beyond question, any one outside of Germany producing or desiring to purchase nitrogenous fertilisers or similar compounds, will have to deal with a single organisation, essentially a branch of the German Government, which will have an absolute monopolistic control of all such products produced in Germany or whatever surplus there may be for export (Hearing before the Committee on Agriculture and Forestry, U.S. Senate, S. 3390, Mar. 22nd, 1920, p. 52)." It is reported that the preliminary allotment of production to the Badische Co. in the Syndicate is three hundred thousand tons per annum, which should leave a considerable exportable surplus. This would constitute a formidable weapon in any price-cutting campaigns entered upon by the I.G. in order to preserve her various monopolies. We learn from the *Colour Trade Journal* of August, 1920, that the German Government has advanced something over ten million pounds for the construction and operation of the Haber plant.

The New German Dye Combine.—Internal changes have accompanied the development of these external relationships. The interchange of capital and directors between the different branches, the use of all assets for a common purpose, and the pooling of all profits effected in 1919, has brought about a closer union. From the relatively loose pre-war combination held together by common price interests, the organisation has passed through the cartel to what is now practically a form of trust. The German dye industry is now a closely woven, almost homogeneous institution. It has added economic cohesion to technical efficiency, and is to-day the largest technically efficient potential instrument of war in the world. We have thus revealed the existence, and indicated the nature, of the resultant activities of the chemical policy guiding the pre-war German combination of organic chemical or dye producers. Further, it is seen how the war stimulated and sealed closer relationships between the constituent firms, and between the resultant organisation, the I.G., and the German Government. Continuing, we find the above tendencies intensified since the Armistice, from unmistakable signs briefly referred to above.

Aggressive Nationalistic Policy.—Both in peace and war, the combination of interests, known as the I.G., has successfully pursued an intensely nationalistic and aggressive chemical policy. We might ignore what some have regarded as the sinister side of the I.G. activities, considering the whole as a wonderful monument to German science, thoroughness and patriotism, which it undoubtedly is in many respects. But the significance to the Allies and associated countries remains the same. Even without any thought or intention on the part of present day Germany to use this thing for war, it remains a serious menace. But the direct evidence which we possess does not actually support such a peaceful view. Her press confidently prophesies the resumption of the pre-war German monopoly, reassuring its readers by careful analysis of the causes of the eventful failure to establish organic chemical industries in Allied countries.

Are we to yield in this field of economic war? If so, then one of the chief lessons of the Great War will remain unheeded, and the future cannot fail to prove this to the hilt, to our cost.

CHAPTER X

The Element of Speculation.—It is of considerable interest to introduce an element of speculation into our discussion of chemical warfare. In glancing at future possibilities, we can adopt one of two courses, follow up the clearly marked lines of recent development, or give the imagination play within the whole field of scientific possibility. The former course lies more within the scope of this book.

Chemical Tactics and Strategy.—Two basic military conceptions come to our assistance in attempting to characterise types of chemical warfare development. With a little explanation it is possible to place this or that method in the tactical or strategic class. Any new chemical warfare development capable, under a given system of individual protection, of successfully attacking the hitherto protected individual, may be termed strategic. The method may be aimed at a protected or hitherto immune human function, but if it overcomes protection it is then capable of effecting strategic results by its use on a sufficiently large scale. Thus we regard the first introduction of cloud gas by Germany, or their use of mustard gas, as examples of strategic chemical warfare moves. Any fundamental discovery of this sort, applicable to chemical warfare, is capable of strategic effects. Used only on a small scale, however, these possibilities may be lost and tactical advantages may alone accrue.

The Riddle of the Rhine

The tactical type of chemical warfare method involves the use of some new or old war chemical device in achieving a tactical objective which may, itself, form part of a larger scheme with strategic significance. Examples were plentiful during the recent war. We may refer to the use of smoke, of gas shell for neutralisation, or of cloud gas as preparation for a local infantry advance.

The same classification can be applied to the protective as to the offensive side of chemical warfare. The equipment of an army of millions with a gas mask has a strategic value, if it counters the large-scale use of gas by the enemy. The mere fact of this protection may serve the same purpose as a violent resistance to a huge enemy attack. It may render the attack, and, therefore, the resistance, out of the question. By permitting the individual soldier to retain the efficient use of his weapons in gas, the mask, or other form of individual protection, may render a costly counter-attack unnecessary. In this way protective methods in chemical warfare may be the determining factor in some strategic campaign or tactical activity. The distinction between tactics and strategy in chemical warfare cannot be made by grouping substances, or their methods of application to war, any more than one can say that certain infantry or artillery formations or weapons have a purely strategic or tactical function. The distinction lies rather in the magnitude and incidence of use of the chemical appliance on the battle-field, while depending on its novel nature. A new substance, possessing potential strategic value, may be wasted, and its surprise effect lost, in some local affair. This applies to the use of mustard gas by the Germans and to our own use of the Livens projector. Our armies were surprised and our plans modified by the

German use of mustard gas at Ypres and Nieuport. We were not clear where this new thing was tending. Think of its strategic and psychological value had it been used on a scale and front twenty times larger. Leaving the chemical field, we can say that the first British use of the tank provided another example.

New War Chemicals.—The question of entirely new war chemicals is of general interest. The first main group of substances with which we were faced during the war contained such types as chlorine and phosgene, whose chief line of attack was directed towards the respiratory system. Specific protection rapidly developed and, once obtained, led to violent attempts to penetrate it or "break it down." In other words, the attempts to penetrate the mask by using higher concentrations of phosgene were analogous, from our point of view, to similar attempts by the use of an entirely new substance aimed again at the respiratory system. The introduction of mustard gas confirmed, what the use of lachrymators had suggested, that the most fruitful line would be found by attacking human functions hitherto immune. First the lungs, then the eyes, then the skin of the human being came under fire, so to speak. What further developments appear possible on these lines? Assuming that means are found to protect satisfactorily the respiratory system, and the eyes, what other vulnerable points can the war chemical find in the human organism? Some more specific vesicant, some modification of mustard gas, might arise, limited in attack to certain portions of the human being. The Germans were already at work on these lines.

"Camouflage" Chemicals.—It is by no means visionary to picture the loss of the sense of taste and smell by the use of some chemical. Partially successful efforts were made by both sides during the war to mask the odour of the harmful constituent of a shell filling by introducing an appropriate "camouflage" compound. Whole series of chemicals were examined from this point of view by the American field laboratory at Puteaux near Paris. The step from specific camouflage compounds to a single general type is by no means unbridgeable in theory.

An insight into work of this kind has been given by Colonel R. F. Bacon of the American Chemical Warfare Service. He says: "The gas-camouflage is of particular interest. It has been found that malodorous compounds (butyl mercaptan, dimethyl tricarbonate, etc.), are useful to mask the presence of other 'gases' or to force the enemy to wear respirators when no other 'gases' are present. As in the case of lachrymators, such 'stink gases' must frequently be accompanied by other 'gases,' in order that the enemy may never know when toxic gases are actually absent. Camouflage gases are also useful in that they save 'mustard gas' and the highly lethal gases. Their value has been demonstrated in trials at Hanlon Field and also at the front." The use of such compounds has an obvious value. By removing the possibility of detecting the dangerous chemical, they enforce the permanent use of the protective appliance or encourage a fatal carelessness in the individual soldier.

Functions Hitherto Immune.—In this field of chemical attack upon hitherto immune human functions, it is particularly easy to class suggestions as visionary and to be wise only after the event. But it must be borne in mind that any nation in a position to effect such a surprise would be in a commanding position. It is believed, for example, that the human being maintains his equilibrium through the proper functioning of the semi-circular canals, organs situated behind the inner ear. It does not appear possible to attain them chemically directly, but they might be reached by the absorption of some suitable chemical into the system in the very small concentrations now possible on the field of battle. We doubt whether any physiologist would go further than to say that such a mode of attack is improbable in the near future. No qualified person would class it as impossible. It has been advanced that the control of equilibrium occurs through the movement of certain hairs through a liquid within these canals. If this be so, then one would simply require to solidify or change the viscosity of this liquid. Would this be difficult? Probably not, for most of the body fluids are of that colloidal nature in which coagulation occurs in the presence of small quantities of special agents. Such a result might cause the individual to lose his equilibrium. This would prohibit all organised movement. An army thus attained would be less mobile than a colony of cripples.

Picture for a moment such a battle as the great German attack of March, 1918—millions of men urged forward from fixed positions under highly centralised control—they advance, say, two or three miles beyond this control and are largely dependent on local initiative for the attack. They then enter clouds of shell chemical and in less than fifteen minutes a fair percentage becomes incapable of advancing in a fixed direction, of obeying local orders, or of anything more than a sort of drunken movement. By this time their supporting artillery would have been identified and attained, and the whole attack reduced to almost farcical conditions. Such a compound may never develop, but who will class it as beyond the realm of eventual possibility?

Every one is acquainted with the peculiar effects produced by various anæsthetics. The emergency uses to which they are put and our personal acquaintance with them may have dulled the imagination. Think for a moment of the possibilities which they unfold. Gaseous anæsthetics, in certain concentrations, produce temporary unconsciousness, other anæsthetics, so called local, produce absolute immobility without loss of consciousness. Chloroform and ether are common forms of the first type, but they are required in such concentrations as to render their battle use impracticable. But the second type, of which stovaine, the new synthetic drug, is a good example, produces its effects in very small concentration. A few drops injected into the spinal column are sufficient to prevent all movement for a number of hours. We cannot expect to obtain the conditions of the operating table on the battle-field, but chemicals which are effective in very small quantities or concentrations may find another channel into the human system. For this reason the development of the mask, the protection of the respiratory channels, is of great importance, for it blocks the way to substances which by mere absorption might produce valuable military results.

Chemical Constitution and Physiological Action. —It is impossible to adopt a more than speculative outlook in this field. So little is known regarding the relationships between chemical constitution and physiological action and very few sound generalisations have been made. A considerable amount of scientific work occurred on these lines in various countries before the war on the connection between the chemical nature of compounds and their taste and smell, but the relationships are still obscure.

Unsolved Problems of Mustard Gas.—The use of a chemical which attacks some unexpected human function introduces many disturbing and disorganising factors. Thus the introduction of mustard gas has left us with a number of unsolved problems. By employing this substance Germany departed from her usual caution and violated one of the first principles of chemical warfare. It is unsound for any nation to introduce a new weapon, unless that nation is, itself, furnished with the means of protection against its eventual employment by the enemy. The Germans have, themselves, explained this breach of the principles of war. They were convinced that we could not retaliate with mustard gas, because we could not produce it. It was a miscalculation but based on grounds of which they were sure, having been largely instrumental in determining them through their aggressive chemical policy.

Mustard gas attacks the respiratory system and the outer skin of man. The armies were efficiently protected against the first line of attack, but they never developed efficient protection against the second. Protection of the skin of the individual soldier against mustard gas was theoretically possible in three ways. In the first place a number of chemical solutions were devised which, applied to the affected skin, would destroy the poisonous chemical. This was a bad method, and was never efficiently employed. German army orders after the French introduction of mustard gas were bristling with references to chloride of lime or bleaching powder. It was to be kept in every conceivable place where the gas was likely to penetrate. Soldiers were provided with boxes of bleach called "Gelbolin." Permanganate of potash was carried as an alternative for a brief period. A wire from the Third German

221

Army to the War Ministry, Berlin, dated 17th July, 1918, stated: "Chloride of lime has all been issued in boxes to the troops. Reserves exhausted." One had the impression of a drowning man catching at a straw. Supply on a sufficient scale to cover most cases was practically impossible. Each soldier would have to carry the protective chemical as part of his equipment, and its proper use depended on training. There was no time to identify and assemble the thousands of affected cases for central treatment. Mustard gas penetrated thick clothing, even boots, and was often only identified hours after the damage was done. The second method which was attempted on a large scale was the protection of each soldier by special mustard-gas-proof clothing, but a man, fighting for his life on the battle-field, will not tolerate such a handicap to movement, and, although hundreds of thousands of oiled suits were prepared and were of definite use in certain special cases, for example in certain artillery formations, yet the method must be rejected as unsuitable from a military point of view. The third solution, which was tried experimentally on a large scale, was to cover soldiers going into action with a cream or paste of protective chemical. This, again, could only be applied in special cases, prior to an assault, for example, and could not be regarded as a permanent form of protection.

As we have seen, mustard gas infected whole areas for many days, owing to its great persistency. It was often necessary to cross such zones for attack or counter-attack. How was this to be effected without huge losses? It was found possible, literally by creating roads of bleach, that is, by sprinkling bleaching powder on chosen lanes through the infected area, to pass columns of troops through such areas, but this cannot be viewed as a practicable solution. Carried to its logical conclusion, it would have taxed the possibilities of supply beyond their utmost capacity. Here, then, we have a case in which it is not possible to protect a soldier by some specific appliance, and the war found us embarking on schemes of protection by the use of chemicals in quantities which threatened to carry us out of the range of possible manufacture.

A New Type of Obstacle.—Chemical warfare has introduced a new type of strategic and tactical obstacle. Mediæval methods of war relied largely on natural and man-built barriers. Rivers, moats, forts were, and still are, to a certain extent, critical factors in war. The conceptions of a Vauban could determine the issue of a campaign. Such obstacles were only effective, however, when properly manned and armed. The Hindenburg Line and the Canal du Nord were tremendous obstacles when backed by German artillery, rifles, and machine-guns, but, without the latter, they would have been mere inconveniences for the passage of an army. The massing of a multitude of guns, used for the first time during the recent war, produced another form of temporary obstacle, but troops could be trained to, and actually did, advance through the barrage. Further, the ultimate limits of supply and the use of counter artillery introduces time and quantitative limitations to the use of the really intensive barrage. Chemical warfare, however, has introduced a method of blocking out chosen areas of the battle-field in such a way as to prevent their effective use for military defence, communications, or other purposes. It is now possible, by chemical means, to give a normal piece of country the same value as a natural obstacle, or one organised for defence by formidable engineering construction, and manned by rifles and machine-guns. This can be achieved by the use of a highly persistent dangerous gas or war chemical of which, so far, mustard gas is the most effective example. We have seen how the Germans formed defensive flanks during their March, 1918, offensive, by spraying certain areas between their fronts of attack with mustard gas. It is true that, in the quantities in which it has, so far, been used, mustard gas has not converted open areas into absolute obstacles against the movement of a determined individual, platoon, or even larger unit. But even in the quantities which have already appeared on the battle-field, it has rendered whole zones practically unusable for huge masses of men, owing to the certainty of a very high percentage of casualties. Up to the present its value has been rather as a serious factor in Staff consideration of losses than as an actual physical barrier. Many of the casualties are only incurred a few hours after contact with the gas. This may not deter a man from crossing an affected zone, but it may deter the Staffs from using that zone, when they realise that this would imply the certainty of many thousands of casualties amongst the troops. The choice is between two evils, tactical acquiescence to the enemy's plan, blocking out a certain area, or the certainty of huge casualties. A very interesting case occurred in the German attack near Mt. Kemmel in the spring of 1918, where large quantities of German mustard gas were used some distance in front of the orginal line of German attack. In this case, not only was it clear that the Germans would not attempt to advance beyond a limited objective (and they did not), but the development of their attack left them organising their defences behind their own mustard gas barrage.

The "Persistent Lethal" Substance.—The importance of these considerations can hardly be exaggerated when we realise that, at any time, a substance possessing the same strategic value as mustard gas, but much more violent casualty effects, may be discovered. The Germans were certainly aware of these possibilities. According to the statement of an apparently reliable prisoner of the 30th R.I.R., July, 1918, the Regimental Gas Officer stated in a lecture that, as the Allies had used a new gas, the Germans were going to employ a "White Cross" gas shell. This gas was "stronger" than any of the gases at present in use; it possessed a persistence up to eight days, and could, therefore, not be used on the front for an assault. Its persistence was favoured by damp or misty weather and by the nature of the ground. Neither the German drum nor the masks of the Allies afforded protection against it. The last important German development consisted in the use of pumice impregnated with phosgene in their Livens bombs. It was clear that the Germans were attempting to produce a gas which was not only highly lethal but persistent. Following up this idea, we can forecast the use of a chemical which will not only permit the formation of defensive flanks, or pockets, in the enemy front, or in our own defensive positions, through their influence on Staff considerations with regard to casualties, but, by replacing the relatively mild casualty effect of mustard gas by a highly and rapidly lethal effect, will render these areas not only strategically, but physically, impassable. One of the most significant possibilities in chemical warfare development is the arrival of this type of the compound, the highly lethal, highly persistent chemical.

The Critical Range.—These considerations are very interesting from the military point of view.

225

The Riddle of the Rhine

Consider the phenomenal amount of muscular energy required to organise any captured stretch of territory against counter-attack. The type of compound we have outlined is likely to change completely the aspect of attack and counter-attack. The Somme battle-field, for example, gave the impression of a series of defensive positions organised by the one side or the other after attack or counter-attack, in order to hold small gains of ground, which were never intended to represent the final advance. Successful progress from one trench system meant building another, under the pounding of the enemy's artillery, and the deadly fire of machine-guns, exposing, in this improvised system, large numbers of troops, among which casualties constituted a continuous drain upon eventual reserves. The arrival of the highly persistent lethal compound should provide an effective substitute for this laborious constructional protection in the shape of the persistent lethal barrage. This will render immediate counter-attack and near machine-gun fire very difficult. Automatically, fewer men will be needed to hold the advanced positions. It is true that, with the next attack, "kicking off" and assembly positions will be required, for these can be much more efficiently developed behind a deep chemical barrage and will demand the exposure of fewer men where more time is available for preparation. Such conditions, however, can only occur if one side possesses some distinct advantage with regard to surprise by, or efficient protection against, the persistent lethal compound. When both sides are equally matched in this respect, a duel will arise in which the winner will be the one who can throw the critical concentrations of chemical into a given area at the greatest range. This might be called the "critical range." Herein lies the importance of the development of such weapons as the

Livens projector, and the Germans had certainly grasped an important principle, when they used our own modified weapon against us with a much greater range than our own. If we admit the possibility of a persistent lethal compound, this question of critical range assumes outstanding importance.

The New No-Man's-Land.—The recent war witnessed a rather sudden adoption of trench warfare, during a period in which the artillery strengths of both sides were relatively feeble, when compared with the later stages of the war. Accordingly, there arose very definite lines of field fortifications, and strongly held trench systems, separated from each other by a comparatively narrow No-Man's-Land. With the development of the formidable artillery strengths of belligerents, there was a tendency to form a much wider No-Man's-Land, and the front line systems were lightly held, approximating, in many cases, to an outpost line.

The discovery and mass production of a persistent lethal substance is likely to convert No-Man's-Land into a permanently infected gas zone, manned by special outposts of permanently protected troops. Combined with the development of smoke, this may render unnecessary the highly organised trench assembly systems of the recent war, used before the assault, and, with the development of the tank as a fast fighting machine, and for the transport of troops, one can obtain a glimpse of the nature of the new attack and counter-attack. A recent writer[1] has shown us the future tank carrying war into the enemy's country and destroying his nerve centres by actually reaching and paralysing the G.H.Q.s. of armies and smaller formations. Such operations will have to occur through a wide zone of the new gas and

[1] *Tanks in the Great War,* Col. J. F. C. Fuller, D.S.O.

227

will necessitate the anti-gas tank. Indeed, one of the most important functions of the tank will be to carry the advance guard of an army beyond the infected No-Man's-Land, and such an advance will occur behind a series of smoke barrages created, in the first place, by the artillery, and, later, by the advance of tanks themselves.

The "Alert Gas Zone."—The development of the "gas alert" idea has definite interest for the future of chemical warfare. It is well known how the development of gas shell and surprise gas shoots by the Germans led to the necessity for "gas alert" conditions between certain times and within certain distances of the front line. The mask had to be worn in the so-called ready position, in order that swift adjustment might be possible in case of surprise attack. The summer of 1917 witnessed a great increase in gas shell activity. This was reflected in important changes in the "gas alert" regulations. In the autumn of that year all periods of readiness were abolished and replaced by a constant state of readiness. In the forward area absolute readiness was required within two miles of the front line, and special precautions were taken as far back as twelve miles. That the Germans suffered under the same restraints is witnessed by many captured documents. In particular, a divisional order taken in December, 1917, gave the gas danger zone as within fifteen kilometres of the front line, and within this region every one must carry a mask. The alert position of the mask was insisted upon within two kilometres of the front line. By July the alert zone had increased in depth in both armies. This tendency must have increased, had the war continued, for both sides were employing gas in guns of larger calibres, and weapons were being devised, such as the improved German Livens projector, which gave high

228

concentrations at much greater distances from the front line, *i.e.* with greater critical ranges.

We have seen how the possible development of a persistent lethal compound may produce an infected and wide No-Man's-Land. Imposed on this, there will, no doubt, be "gas alert" conditions of great depths. How do these conceptions work out for the war of movement? It would appear that the possession of such a compound and the means of producing and using it on a very large scale could determine the stationary or open nature of warfare, if other forces were not too unequal. A new military factor emerges, the artificial, permanent, unmanned obstacle, which can be laid down at will on areas whose magnitude depends finally on manufacturing capacity. The germ of the idea appeared during the war at Kemmel and in the various mustard gas barrages formed by artillery or delayed mines used by the Germans in their great retreat. The sudden development of such barriers will be equivalent in effect to the creation of strong trench systems, but these could never result, under war conditions, in time to approach the strategic flexibility and importance of the persistent lethal infected barrier.

Gas and Aircraft.—The combination of gas and aircraft presents the possibility of attaining strategic effects by chemical means. Many rumours were afloat, towards the end of the war, regarding the use of gas by enemy aircraft, and there was apprehension amongst the civil populations, which has been reflected in numerous public utterances. Evidence on the matter is very scanty. In July, 1917, the use of gas in aeroplane bombs by the Germans was reported, but not confirmed. Further reports in August indicated the use of Blue Cross, owing to the sneezing effects which were produced on those within reach of the air bomb. In October, the evidence was more conclusive. But the German aeroplanes left no blind or dud shell, and, beyond the violent nasal and sneezing effects of Blue Cross, evidence was again absent. This report was very persistent, for, in July, 1918, there were again rumours that Blue Cross bombs had been dropped on the British near Ficheux. The Air Forces of the different armies were, perhaps, the last to feel the effects of the gas campaign, but the pilots of low-flying aeroplanes in the 1918 offensive were constantly crossing pockets of gas, and this, added to the fact that the pilots were often compelled to land in gas, led to their equipment with gas masks. A respirator of special type was taken from a German aviator in April, 1918, after the fighting at Passchendaele. But the war gave us no direct evidence of the successful use of gas and war chemicals from aircraft. This, however, is no criterion as to its eventual importance. The Allies definitely refrained from employing the combination until Germany should give them the start in what was regarded as a new atrocity. The main reason for their lack of development on these lines was probably the fact that the most suitable type of gas only developed during the later stages of the war, when it was required exceedingly urgently on the front. No really harmful persistent compound appeared before the advent of mustard gas, and the dangerous non-persistent types, such as phosgene, could not have been used with great success, owing to the fact that very considerable quantities would have been required to produce any serious effect. Mustard gas, however, which could have haunted a city for days, would not have been required in such large quantities. But its more urgent need on the front, and the fact that soon after it arrived the Germans were sending out feelers to see whether the Allies would consider the cessation of chemical warfare, were probably sufficient reasons to explain their failure to use it from aeroplanes.

Another point must be raised in connection with the use of gas from aircraft which has not yet received much attention. We must remember that the use of projectiles from aircraft over a city was a very different proposition from their use over a battle-field. One of the advantages of gas over explosives on the field of battle was its greater range of action. It produced effects at longer distances from the point of impact, but no such incentive existed for the use of gas from aeroplanes over large cities. Explosives, which might miss their objective on the field of battle, could not do so in a city. They were bound to hit something. The load of the aeroplane is always important, and the essential is to carry, weight for weight, the material which will produce the most effect. There is no doubt what this will be when the persistent lethal compound arrives, and mustard gas would probably have been superior to explosives for use by German aircraft on British cities.

Protective Development; — Individual Protection.—The question of protection against chemical attack presents some knotty problems for the future. Let us glance at the broad lines of war development in this field and forecast their future in a speculative way. Protection developed along two main lines. Individual protection covered the mask and any other protective appliance used by the individual soldier, while the term collective protection was applied to any method or appliance which afforded simultaneous protection for a number of individuals.

In general, the former represented an attempt to purify the poisoned air actually inspired by the soldier, whereas the latter was an attempt to purify

231

the atmosphere of a locality or to prevent its initial poisoning. How far can the individual form of protection develop to meet the possibilities of the chemical attack? It certainly seems to have countered satisfactorily all the war attacks upon the respiratory system, although, as we have pointed out, the Germans might have failed, had we been sufficiently prompt in introducing our arsenic compounds. But we have forecasted the use of chemicals which may attack human functions hitherto immune. For the sake of our argument, we can divide these into two classes, those attained through the respiratory and digestive systems and those attained through contact with some other part of the body. The former can probably be satisfactorily met by developments in the mask. Even that does not appear certain, when we remember the emphasis laid by Germany upon the possibility of penetrating the mask by using a particulate cloud. The last word has certainly not been spoken in the struggle between the mask and the chemical attempting to penetrate it. But both the introduction of mustard gas and general speculative grounds justify us in concluding that attacks may materialise upon other parts of the human organism. We cannot foresee the actual point of attack and can, therefore, only view with assurance some form of protection which covers the whole body.

Collective Protection.—All parties dabbled in such a form of protection, but the French were the only ones to make a large-scale experiment on the front. It was not very successful, for the burden of these oilskin suits was intolerable. It may be that some successful form of protection for the whole body will materialise, but on general grounds we can assume that development will follow other lines. What are the possibilities? They all lie in the direction of collective protection. The individual cannot be satisfactorily protected from the new gas and remain an efficient soldier. We must, therefore, see whether it is not possible to protect numbers of men by removing them from contact with the poisoned atmosphere. A stationary form of such protection was used by all the armies, but emphasised by the French, by the creation of a large number of enormous underground chambers, some capable of holding more than a thousand men, the entries to which were carefully protected by special filtering devices to prevent the entry of the poisoned external atmosphere. On the British front these enormous dug-outs, although not absent, were largely replaced by the efficient gas-proof organisation of the smaller dugouts. The use of impregnated blankets for this purpose must be well known to any who visited the front or took part in hostilities. But you cannot imprison a whole army in this way. The value of these collective protective chambers depended on the fact that a certain number of men were always on the alert in the defensive systems outside and around the chambers, exposed to those gases against which the latter chambers were devised.

In my opinion, the further intensive development of gas warfare, such as would have accompanied, say, the doubling or quadrupling of the German factory output, would have forced us into realising the limit of this collective protection. It would have compelled us to immobilise, in these shelters, more men than was consistent with the safety of the zone in question. Undoubtedly, the future of collective protection lies in some form which will leave the soldier his combatant powers, in other words some mobile form. This has already been forecasted by Colonel Fuller in his book on *Tanks in the Great War*. But

233

he passes lightly over the protection of the tank against gas. With the increase in depth of infected zones, through the increasingly lethal nature of the persistent compound, the tank will be compelled to rely on filtration methods of protection, instead of the use of compressed oxygen in a gas-tight compartment. Once committed to the use of oxygen, the only safe procedure will be to close up the tank and employ the oxygen while there is any suspicion of the presence of gas, and, under these conditions, oxygen transport would become a factor militating against the prime purpose of the tank, the transport of troops and arms. It is safe to forecast a tense struggle between chemical weapons and protective tank devices in the event of future wars.

Conclusion.—The facts which we have surveyed in early chapters, and the development foreshadowed above, form part of a much wider subject, for they are but one aspect of scientific warfare. In what main directions has science modified or revolutionised modern war? Its influence has touched practically every weapon in manufacture or design, introducing profound modifications in many cases. The sum total of such changes may be claimed to have revolutionised warfare, but the term revolution should be reserved for some more specific scientific innovation, which threatens to change the nature of war rather than merely improve existing weapons. Modern wars have all echoed the popular cry for some new scientific principle or device to settle hostilities with one sharp stroke. This conception has been the sport of writers of fiction and others for many years. The "electric" death-dealing ray, the all-powerful gas, the deadly bacteria, and the "explosion" wave have all shared in buoying up the hopes or quickening the fears of warring peoples. Contrary to popular supposition, a decisive scientific military surprise of this nature is not likely to follow close on the heels of the discovery of a new phenomenon. It is more than eighty years since the mind of a Faraday delved so fruitfully into electrical science, yet the oft prophesied large scale direct use of high voltage electricity, or some other form in war has not materialised. Organic chemistry was a well-founded branch of science early in the nineteenth century, and flourishing industries, fostered by it, were in existence thirty years ago, yet it was not until the early twentieth century, and the recent war, that we witnessed the rapid growth of organic chemical warfare, which, I claim, was as revolutionary as any other war development. The physical sciences have left their mark on every weapon and mechanical appliance, and the cumulative effect of these changes is indeed large, but the most revolutionary upheaval in warfare, with permanent results, came from chemistry. The flexible nature of organic chemistry must not be lost sight of. In the physical sciences, electricity, for example, years of co-ordinated world progress are required to produce an epoch-making discovery which might have critical and direct war significance. Radioactivity has shown us what undreamt-of energy is bound up in the atom, and many are the prophesies regarding the harnessing of these forces for constructive activities. At least one prominent novelist has pictured their destructive use in the radioactive bomb. But the use of this wonderful store of energy for peace and war can only result from years of costly and voluminous research, and we have no idea of the difficulties involved in production, without which any invention, however telling and revolutionary, has no incidence on war. But in organic chemistry a single worker, following up some rare family of compounds, may stumble upon a sub-

235

stance not far removed chemically from related compounds yet infinitely more potent for war. Mustard gas, or B:B dichlordiethylsulphide, is a member of a group of compounds differing only slightly in chemical structure the one from the other. Yet its nearest chemical relative is comparatively harmless. The persistent lethal compound which will vastly change the nature of warfare will probably be but a slight chemical modification of some harmless substance. Thus, by comparison with other branches of science as the handmaids of war, organic chemistry is sympathetic, flexible, and theoretically capable of yielding revolutionary discoveries in a relatively short time. We can only base such speculations on general grounds. Circumstances may disprove our contention over a short historical period; but it will be borne out in the long run. This is not the only reason, however, for the unique war importance of organic chemicals. It so happens that many of them are essential to our daily life, as dyes, drugs, photographic and other synthetic products. Industries, therefore, have arisen for their manufacture. And this is not all. Organic chemical factories have proved to be not only arsenals in disguise but endowed with the flexibility of their parent, the science itself. The factories and plants ignore the war significance of the problems put to them. They can develop the production of practically any chemical which research can produce. The will of man can thus silently and swiftly convert the dye factory into an arsenal.

These inherent possibilities of organic chemistry, flexibility in research and production, make chemical warfare the most important war problem in the future reconstruction of the world.

CHAPTER XI

HUMANE OR INHUMANE?

A good deal of abuse has been showered on chemical warfare methods by those who understand very little about them. It has been claimed by such that gas is particularly atrocious. Feeling on the matter has been so strong in certain quarters that the fact that all war is particularly vile and atrocious seems to have been completely lost sight of. Let us take up this matter in a rational way. In the first place, what do we mean by the atrocity or inhumanity of a weapon? We can either appeal to the imagination or the reason, in the first case, by visualising the battle-fields, or, in the second, by making a cold analysis of the casualties caused by gas.

Nature of Gas Casualties.—Every normal person who experienced and survived the throes of the different stages of the war, and of the different gas surprises, mainly German, which were sprung upon us, finds it difficult to think out, or express, a cool and balanced view on the question of poison gas. But such a balanced view is most important for the future. It must be remembered that the official protests in 1915 arose on the grounds, to use Lord Kitchener's words, that "they employed these poisonous methods to prevail when their attack, according to the rules of war, might have otherwise failed." Had the rules of war permitted their use, we should, no doubt, have been protected. But these protests, submerged in

The Riddle of the Rhine

popular sentiment, became an outcry against the atrocity of the new weapon. This, a just criticism at the time, became inaccurate when the Allies reacted, methods of protection developed, and the specific tactical uses of gas were realised. The view of the peculiar atrocity of gas has outlived the truth of war experience with regard to it. We agree that chemical warfare is atrocious. But it is no exception, for thus are all the aggressive methods of warfare. Indeed, when we attempt to interpret atrocity in terms of available casualty statistics, we find that gas is slightly less atrocious than the other weapons. We must either incline to this view or dispute the figures, which are authoritative. Consider the American figures. These will be more truly representative than our own, because their troops were only exposed during the later and more developed phases of the war. Of the total strength of the A.E.F., the number gassed was about six per çent., wounded by rifle and machine-gun fire about one per cent., wounded by high explosive one and a half per cent., shrapnel wounds three per cent., and bayonet wounds less than one half per cent. But although enemy gas caused more than 70,000 casualties, yet of these only one and a half per cent. were fatal, while the total number of deaths for all types of casualties was thirty per cent. Thus against the American army, measured by casualties produced, gas was by far the most effective, and yet by far the least deadly weapon. What can be more atrocious than the actual cone of tens or even hundreds of dead and wounded invariably left before an untouched machine-gun emplacement in an assault? What is more horrible than the captured first line trench after its treatment by the preparatory bombardment, or the mutilation of men peacefully sleeping in billets behind the battle front and thrown, broken and bloody,

238

through their billet walls under the wheels of passing transport, as one has seen them?

The whole experience of real war is beyond adjectives. But, leaving impressions, let us turn to facts. With regard to the future and from the point of view of atrocity, gas has a hopeful outlook as compared with other weapons. This may seem a curious statement to make, but consider the following. We cannot envisage advances in the use of explosives in shell or bomb to render them more humane. Explosives, if their development be pressed, can only become more violent, with a wider range of action. Chemical warfare may follow the same lines, but it has the unique possibility of developing on more humane lines. The vesicant action of mustard gas produced huge casualties with relatively little permanent harm. Chemicals may be found which temporarily influence human functions, enabling military objectives to be attained with a remarkably small amount of pain and death. In a fair review of the whole situation, this possibility cannot be overlooked. It is more than possible that a League of Nations, compelled to employ an element of force in its eventual control of peace, may find its most effective and humane weapon in some chemical development. However visionary these views may appear, they are not unjustified as scientific possibilities. Analysis of war gas casualties reveal two main trends. As the struggle became more intense the number of casualties multiplied. They were considerable during the first period of cylinder attack, and the rate remained steady until the beginning of the mustard gas period. From the summer of 1917 to November, 1918, there were more than ten times as many gas casualties as for the preceding three years of war. But the percentage mortality, the number of deaths amongst each hundred men attained, decreased considerably. As high as twenty-five per cent. during the early cylinder attacks, it decreased to two and a half per cent. for the huge number of mustard gas cases. Yet mustard gas was an exceedingly important military factor. It illustrates the possibility of development on these lines, but we must by no means disregard the atrocity of chemical warfare, and safeguards are required for the future.

We cannot do better than conclude by quoting from General Hartley's report to the British Association. He says:

"The general impression that gas is an inhumane weapon is derived partly from the German breach of faith in using it contrary to the Hague Convention, and partly from the nature and number of casualties in the earliest cloud attacks which were made against unprotected troops. Under the stress of a long war the individual is apt to forget the physical and mental sufferings it involves, unless he is daily in contact with them, but a dramatic occurrence such as that of the first gas attack forces on the imagination the brutal significance of war—the struggle for victory by killing—and the new weapon is judged as inhumane, like gunpowder in the fifteenth century. If we accept war as a possibility, the most humane weapon is that which leads to a decision with the smallest amount of human suffering and death. Judged from this standpoint, gas compares favourably with other weapons during the period when both sides were fully equipped for offence and defence. The death-rate among gas casualties was much lower than that among casualties from other causes, and not only was the death-rate lower, but a much smaller proportion of the injured suffered any permanent disability. There is no comparison between the permanent damage caused by gas, and the suffering caused to those who were maimed and blinded by shell and rifle fire. It is now generally admitted that in the later stages of the war many military objects could be attained with less suffering by using gas than by any other means.

Sargent's Picture.—"The judgment of future generations on the use of gas may well be influenced by the pathetic appeal of Sargent's picture of the first 'Mustard Gas' casualties at Ypres, but it must not be forgotten in looking at that picture that 75 per cent. of the blinded men he drew were fit for duty within three months, and that had their limbs and nerves been shattered by the effects of high explosive, their fate would have been infinitely worse."

Need for Safeguards.—We have continually referred to the need for safeguards instead of mere reliance on prohibition. Such views and facts as the above should be more generally known in order that very worthy sentiments may not impel us to adopt an unsound solution for future peace. However alarmed and revolted we may have been in 1915 and later during the war, it is essential to take a balanced view in the present critical period of reconstruction.

CHAPTER XII

CHEMICAL WARFARE AND DISARMAMENT

Preceding chapters have shown how chemical warfare has now become a normal, technical, and increasingly important part of the science of war. Further, it has opened vast possibilities, the limits of which it is very difficult to fix.

The Treaty of Versailles.—Chemical warfare received definite attention in the formulation of the Treaty of Versailles. Lord Moulton, one of the few Allied representatives who realised the full importance of the matter, has drawn attention to its Treaty aspect in a recent speech. He lays emphasis on the fact that the full significance of the German dye industry was not realised during the war. Referring to its chameleon-like nature in peace and war, Lord Moulton says: "All this was imperfectly present to my mind throughout the war, and I was aware of the gravity of the matter, but until I learnt what had passed in Germany I could not appreciate it fully. I have spoken to you of the extent to which the Germans turned their chemical works into general works for supplying explosives. I have not touched the part in which they played the most deadly game against us, and that was where they used their chemical works to produce those toxic gases."

The same statement tells us, "The knowledge that I have gleaned as to what was going on in Germany during the war makes me feel that all my anticipations of the importance of chemical industries in time of war, all the views that I expressed of that importance, did not nearly approach what has been proved to have gone on in the enemy's country during the war." He then proceeds to explain how a clause was inserted in the treaty—"whereby the Germans have to tell us all the secrets of their manufacture of explosives, all their methods of making toxic gases— in fact, all the military secrets that made them so terrible. This clause was a very just one. It is not fair that when we have gone through this agonising struggle, and when we are still suffering from the consequences of all the wealth of knowledge and ingenuity which they employed for their infamous purposes—it is not fair, I say, to allow them to keep these secrets to themselves, and I think you will agree with me it was in the highest degree consonant with justice that we should make them reveal them all to us." Small wonder that we missed this vital point, that we failed to fathom the force behind the German chemical war, if such an eminent authority was left groping for the truth. There was no time for mature reflection with the problems of war supply pressing forward in an endless stream. Lord Moulton was himself responsible for the brilliant solution of the most important, the problem of explosives supply.

The realisation of the facts in question led to the direct admission of their importance in the Treaty. Article 172, the one in question, states: "Within a period of three months from the coming into force of the present Treaty, the German Government will disclose . . . the nature and mode of manufacture of all explosives, toxic substances or other like chemical preparations used by them in the war, or prepared by them for the purpose of being so used."

German Information.—This clause should be fulfilled in detail. In any given period of the stage of intensive chemical warfare and at the end, the Germans, in addition to those devices in operation, must have had a large number of more telling and more novel ones in preparation. It is important to get as much information as possible on this development.

A striking fact emerges. The years 1915, 1916, and the early part of 1917 witnessed the actual manufacture of the war chemicals which were used by Germany on the front. All the research and other work which precedes chemical manufacture must have been completed much earlier. What surprises, then, had the German laboratories in store for us after 1917? Have these been revealed under authority of the Treaty?

Probably the most important point in the clause is its interpretation with regard to the Haber process. Its critical importance in the manufacture of explosives is so great that our neglect to use the Treaty to remove the monopoly is a direct menace to peace. This process undoubtedly saved Germany in 1915 and is largely responsible for the three years of war agony which followed. It can only have missed specific reference in the Treaty on account of its claim to represent the fertiliser rather than the explosives industry. To yield to such views, however ideal the motives, is to threaten the greater ideal of world peace.

Limitation of Armament.—This clause, covering only war development, cannot be regarded as a serious safeguard for the future. It is rather the fruits of victory, the logical outcome of Allied success and the German breach of faith. But the Treaty of Versailles contains an admission of the importance of chemical warfare for the future. Article 171 states: "The use of asphyxiating, poisonous, or other gases

244

and of analogous liquids, materials, or devices being prohibited, their manufacture and importation are strictly forbidden in Germany. The same applies to materials specially intended for the manufacture, storage, and the use of the same products or devices." What kind of guarantee is this? How far is it supported by other disarmament? It is very important to answer these questions. In a sense the full execution of the other relevant Treaty clauses would provide a partial answer. We deal with these in the next chapter.

Report of the Hartley Mission.—Chemical warfare is the *point faible* in world disarmament. Judging from the above clause of the Treaty, it is clear that this is not fully recognised. Once again our trust is invited in mere prohibition. The lesson of the war is not learnt. The chemical menace is not countered. Why should this be? There are two main reasons. In the first place, very few had any conception of the tremendous growth in this branch of warfare, for facts had rarely been disclosed, and those with no direct contact with chemical warfare were relying on impressions. The vivid recollection of the first German cloud attack, and of the introduction of mustard gas, have, for most people, obscured the solid facts of the case. The great importance of the projector, the high percentage of chemical shell used by the enemy artillery, and the tremendous undertaking involved in protecting an army of millions with a modern gas mask, have not been grasped. The Hartley report clearly revealed the importance of the German dye factories for chemical warfare production. But we have a shrewd idea that it left many of its official readers much better informed on production than on the use of the materials concerned, that is, on the military value of chemical warfare.

245

New Conceptions in Chemical Disarmament.—
The second difficulty preventing a full understanding
of the case lies in the fact that chemical disarmament
involves certain conceptions which are remote from
the normal military outlook. Let us examine the
matter as simply as possible.

During the many discussions on disarmament in
Paris, various principles were suggested as a basis.
One which received recognition in the Treaty was the
limitation of the number of projectors or guns, using
the term "projector" in a general way to cover all
projectile-throwing weapons. Thus, in the sense implied, rifles, machine-guns, field and heavy guns are
projectors. Recent writers have termed gas a projectile, one which, on account of its fluid nature,
ignores the limitations of explosive shell and multiplies
their radius of action indefinitely. This is true—with
one most important qualification. Gas has never entirely depended upon the usual form of projector, the
gun, and with the limitation of the latter its dependence will decrease. New forms of chemical weapon
will evolve. Now it is true that almost every form
of warfare which one can conceive depends for success on some sort of projector, and it is also true that
the manufacture of these projectors can be controlled,
because it is usually so complicated. These remarks
apply, for example, to the manufacture of a field or
heavy gun. But there is one serious exception to the
covering power of this method of limitation. You
cannot carry on tank warfare without ordinary projectors, but you can run a chemical campaign without
them.

Facing the difficulties which are before any League
of Nations or international body planning world disarmament, let us assume armament reduced to a
police basis. In other words, the use of force is not

entirely ruled out, but is limited to the minimum required for reducing local disorder, maintaining the peace, and contributing to any general scheme for preventing war. The nations, then, agree to limit their personnel and material within certain prescribed bounds. The work of the League of Nations, or central organisation, does not finish here. We cannot assume that permanent purity of national intentions, in other words, some check or guarantee must be instituted. This may take the simple form of systematic reporting by nations and their inspection by the League. Here we meet with considerable difficulty. Unless some simple covering principle for inspection can be determined upon, we shall end up with one-half the world inspecting the administration and organisation of the other. The matter becomes an absurdity.

Limitation, Mechanical and Chemical.—Considering the present trend of war development, we can divide the factors requiring limitation into three classes—the combatants, and weapons of a mechanical and chemical nature.

Tank Disarmament.—A little thought will show that the limitation of the number of projectile-throwing weapons covers the first two types, and is a matter which is not theoretically beyond the possibility of inspection. Periodic inspection could reasonably be regarded as a check against very big scale production beyond the normal scope of industry, for such weapons as rifles, machine-guns, field and heavy guns. If we consider the most important new mechanical war appliance, the tank, we find it no exception to the above remarks. Without projectors, that is, machine-guns, rifles, etc., it merely becomes a means of conveying troops and material from one place to another.

Two possibilities then arise. The number of tanks

247

required might be so small that they could be suitably armed with light projectors without entering upon large-scale production. Secondly, the tank might become an offensive weapon without projectors, by the use of some chemical contrivance. This merely goes to prove that steps must be taken to limit the output of the tank itself. Are such steps possible? We assume that the modern tank is, and will increasingly become, a weapon practically as specific as a big gun, requiring a number of special parts which normal industry does not provide, and that the production concerned can be controlled by inspection with the same order of difficulty as that of the bigger projectors. We now come to the third type requiring limitation under a disarmament scheme.

Chemical Limitation.—Can we limit chemical armament? Our review of production has shown the impossibility of doing so, unless we completely wipe out the organic chemical industry which is essential for world progress by its contribution of dyes, drugs, and other synthetic commodities. The factories of the organic chemical industries are more silently converted into arsenals than any other type. It is true that, under normal conditions of warfare, the decisive success of a chemical campaign might be restricted by the use of other weapons, such as artillery. But, under conditions where the latter are seriously limited, the chemical weapon becomes, relatively, of much greater importance. One of the main trends in chemical warfare was the development of devices which would give long-range chemical effects without a complicated form of projector, or with none at all. Having thus shown the independence of the chemical weapon, under conditions of limitation of armament, we are faced with an important question. What can be the guarantees for the limitation of chemical warfare?

Research.—In the first place, can any research results accrue under Treaty or League conditions? The chief poison gases used during the war owed their discovery, as individuals, to pre-war research which was not stimulated by the need for an offensive chemical. Phosgene was discovered in 1811 by J. Davy, while experimenting on the action of sunlight on a mixture of carbon monoxide and chlorine. Guthrie, in 1860, trying to throw light on some theoretical aspects of organic chemistry, examining the nature of certain so-called radicles or groups of atoms, came across a family of compounds of which mustard gas, or B: B dichlordiethylsulphide, was a member. This he found to be a dangerous substance, but the nearest members of the series were harmless.

These substances will arise as a result of normal chemical research. We admit they may multiply much more quickly if work is specially directed towards their discovery, but it is practically impossible to control such work. The research worker's nearest confidante and laboratory companion might be unaware that he was developing some new vitally important chemical for warfare. No serious person can claim the possibility of a check upon such research. If, then, the Government of any country desires to provide its chemical factories with suitable subjects for chemical warfare production, these can be produced under *any* international arrangements, however prohibitive.

Production.—But what of production? Here, again, we have an entirely different problem compared with that of limiting the output of a gun. Let us assume that the production of some vitally important new organic compound involves four different steps, and that the last step produces the toxic substance. This is a fair assumption. Let us further assume the most favourable condition for detection, *i.e.* the final product is a liquid or gas with obviously toxic properties. Given a big organic chemical industry, there is no possibility of detection by open methods of control. With regard to the first three steps, in practically every case they will be related to some new or existing dye, drug, photographic, or other commercial organic product. The products of these first reactions can either be stored, ready for the rapid realisation of the last reaction, in which case there is no possibility of detection, or the reaction can be completed and the materials passed without exposure through a standard type of plant to an easily concealed container. The only type of inspection which could possibly cope with such a problem would require to probe deeply into the technical and commercial secrets of the factories and plants, and could even then be misled owing to the constantly developing nature of the compounds produced. The inspectors would require to be numerous and as closely in touch with the plants and processes as the actual factory staffs.

Consider the Leverkusen works for a moment. They cover a very wide range of products, are admirably planned on a well thought out and rational scheme, and there is a reason for the position of every unit. Their methodical arrangement would be of more assistance to inspection in this than in any other large organic chemical works with which we are acquainted. Even under such favourable conditions satisfactory inspection would be most difficult. Each one of the twenty huge blocks contains many units of plant, and is devoted to the production of primary, intermediate or finished materials. For the inspection of suspected poison gas production, an examination of the first two would be of no assistance, for the war and peace materials would be identical. Differentiation would occur in the dye and finished product blocks. Each one of these blocks may be producing as many as one hundred different compounds at the same time, and each one of these compounds may, itself, involve two, three, or four different stages. The members of one official mission, when asking to be shown the plant for the manufacture of p-amidophenol, an important dye and photographic chemical, were taken to a large building filled with assorted plant, and were told by the guides, "We have no special plant for the product you mention; we make it in this building with a great many other products, for it is our principle not to have plant which makes one product only, but is readily adaptable for making a variety." In many of the processes the materials do not appear to the naked eye after their introduction into the first plant unit, being fed by gravity or pressure from one enclosed apparatus to another. It would be absolutely essential for any inspection to conduct chemical tests at the different stages. The difficulty of inspection is incontestable. It could be done with a large staff, but we must remember that the Rhine plants are, themselves, but a small corner of the whole world of industry requiring inspection. Even under the most favourable conditions for detection, the chances are exceedingly small. But, in most cases, an enemy with a strong organic chemical industry need not undertake manufacture during peace. He could rely on the potentialities of his chemical industry, which would enable him to commence production in his existing plant immediately on the outbreak of war. The question of the use of the chemical then arises. If of an exceedingly novel and decisive nature, it could take its share of use in the limited number of guns available; on the other hand, it might be capable of use in one of the very simple weapons already devised for chemicals, or to be devised in the future.

Consider the Livens projector, by no means a favourable case. The latest German designs have a range well over a mile. This range may be increased. Yet the Livens projector can be made without serious or obvious war modification of plant, in a tube works, where the bomb can also be produced. The very nature of chemical warfare is such that great accuracy is not required, and simplification of production of the gas projector follows naturally. We conclude from the above that whatever treaty or international arrangements exist for prohibiting chemical warfare, we can find no safeguard in practicable methods of control, and must find safety in some other measure.

Mechanical and Chemical Preparations for War. —There is a fundamental difference in preparation for the mechanical and chemical methods of war. This difference necessitates special consideration for the chemical method from the point of view of disarmament. All the modern mechanical types of war appliances are characterised by their great structural intricacy, witness the Lewis gun with its innumerable complicated parts, the heavy and field guns with their wonderful mechanism, and the future tank with its anti-gas, anti-water, and general anti devices. This characteristic of great structural development has certain concomitants which are of considerable military importance. It imposes certain conditions on production, involving special factories for special parts and other factories for the assembly of those parts. It implies large scale experimentation for the improvement of the appliance. All this brings control and inspection within the region of the theoretically possible, and militates against sudden surprise. The structural characteristic also imposes certain important conditions in military training. It takes a definite period of time to create a machine-gunner who will humour the wonderful mechanism which he serves. He must know the different jambs, and simple repairs. He must be trained. The same remarks apply to any other structurally intricate appliance, such as the tank. In other words, this characteristic is a distinct check on any nation aiming at a sudden expansion from limited to war armament.

But consider the chemical method. The specific property of the chemical which gives it its military value is ultimately its influence on the human organism, which causes casualties or imposes heavy military handicaps on protected troops. There is, again, a question of structure, the chemical structure of the substance in question. This, however, does not involve the same aids to armament limitation as for the mechanical type, unless it be in a very restricted sense. In research, the discovery of the most effective chemical the world will ever see can occur by the use of a few beakers, pots and pans, and common chemicals, directed by a trained mind. Being atomic or molecular, the structure imposes no large scale conditions on the research. Nor is it fair to say that from the point of view of production there is a parallel between the complexity of the molecule and the plant required to make it. The chemically complicated Blue Cross arsenic compounds were produced by Germany in a plant which was simplicity itself when compared with the marvellous installation developed to produce oleum, a concentrated form of the relatively simple sulphuric acid, a fundamental substance in explosives production. Instead of manipulating a huge lathe, or forge, or exceedingly complicated multiple mechanical device, you manipulate temperatures and pressures and vary the reaction medium. Naturally, chemical engineering is very important, but its magnitude and complexity is in no sense parallel with the intricacy of the chemical molecule, whereas a distinct parallel exists for the mechanical war appliance. More than this, we believe that developments in both fields will exaggerate rather than diminish the difference. We see thus how, on general grounds, the chemical weapon tends to evade any normal condition of limitation which might be perfectly adequate for the mechanical type.

Recent Disarmament Proposals.—A superficial examination of recent disarmament speeches by prominent League of Nations advocates leaves one with the glow of inspiration produced by homage to a great ideal. But later reflection, in the cold light of reason, produces a critical, but not cynical, frame of mind. Disarmament depends for success on the way in which we tackle certain critical cases. The carrying out of the more commonly considered forms of disarmament will give immensely added importance to other forms of warfare which have already challenged supremacy in the keen competitive atmosphere of the great world war. The outstanding example is the chemical arm, whose peculiar requirements in any scheme of disarmament have been but vaguely understood.

The great ease and rapidity with which the German dye factories mobilised for poison gas production on a super-industrial scale has already been demonstrated. It took forty years and more to develop those factories. Yet forty days saw many of their plants producing huge tonnages of poison gas, and as many hours were sufficient for others. In some cases, indeed, they were already producing eventual muni-

254

tions long before the outbreak of war. We must not remain insensible to the double-edged nature of this industrial weapon. When with one hand Germany withdrew life-giving drugs from America, with the other she poured upon us an endless stream of deadly poison made in the same factories. Even when our textile industries were threatened through lack of indigo, from the very plants on which we had depended there issued a steady stream of mustard gas, each ounce of which threatened Allied limb and life. But how does this touch disarmament? Very simply. A few quotations from some recently published disarmament utterances will show that we are not pressing the point without need. But let us follow the matter through in a logical way.

The Covenant of the League;—Need for Guarantees.—We start from the sure ground of the Covenant of the League of Nations. Article 8, recognising the reduction of armaments to the lowest point consistent with national safety, refers to the formulation and revision of plans for such reduction and states: "The members of the League undertake to interchange full and frank information as to the scale of their armaments, their military and naval programmes, and the conditions of such of their industries as are adaptable to warlike purposes." Here is the frank admission of the importance of such industries. But later exponents of the League express dissatisfaction with Article 8, claiming the wording to be vague. Thus, from Major David Davies, M.P.,[1] "The whole wording of Article 8 is vague. These proposals would not eradicate the old atmosphere of suspicion which has brought about so many wars. Nations who put their trust in the League are entitled

[1] *The Flaw in the Covenant and the Remedy.* Major David Davies, M.P.

255

to an assurance that the League will be able to enforce its decisions with promptitude. The proposals concerning armaments in Article 8 and elsewhere do not give this assurance. Something more definite is required," and he proceeds to lay down three aims which must be covered by an efficient disarmament scheme.

> "(*a*) Allow each nation an army sufficient to maintain internal order within its own boundaries, and sufficient also to furnish its quota for the League of Nations when required.
> "(*b*) Ensure that the quota of any nation shall not be rendered useless by the employment of a new weapon of war by another nation.
> "(*c*) Provide the League of Nations with an adequate force for immediate use.

"All the above essentials are incorporated in a scheme for an International Police Force. This scheme, which is given in the merest outline, is based on the assumption that our national security must always be absolutely safeguarded, and that before we decide on any relaxation of our armament policy we must be certain that the alternative offers complete protection." Other exponents emphasise this last essential. This reference to an International Police Force raises an important issue. Such a force must draw its personnel from the different nations. Without any doubt, one of the most important contributions from the nations is the fostering of organic chemical research and technical cadres which can only be maintained under true disarmament conditions by the redistributed organic chemical industries.

Viscount Grey—"Germany Must Disarm First." —Viscount Grey, at the public meeting in support of the League of Free Nations on October 10th, 1918, stated: "Germany must disarm first. She led the way up the hill in increasing expenditure on armaments. She must lead the way down the hill. That as a first condition, from our point of view, goes without saying. There can be no talk of disarmament until Germany, as the greater armer, is disarmed." One can only heartily agree with such expressions, but the *dénouement* brings a sense of disappointment. There is a feeling that those who should be nearest are but groping for a solution. The peculiar significance of chemical warfare for the future is freely admitted in these utterances. Thus Major David Davies states: "If they had kept their intentions secret until they could utilise a thoroughly deadly gas in the general attack, it was more than possible that they would have completely broken the Allied line," and Lord Grey, "You cannot limit the amount of merchant ships or commercial aeroplanes, and the fewer the armaments, fighting aeroplanes, and ships of war, the more potential as weapons of war become the things which you use in commerce—ships, aeroplanes, chemicals of all kinds."

Left in this state the case is true but not complete. The essential point is that the new and telling types of armament will develop from these very peace industries. We are not merely concerned with their relative magnitude in a state of disarmament, but with the critical types which may develop from them.

So far, so good, but what steps are proposed to counter the menace? In reviewing what has been suggested by different responsible individuals, we find that the methods intended to cover armament limitation for the newer weapons fall into two classes.

Suggested Methods.—In the first place, it is suggested "that war's newest weapons—poison gas, aeroplanes, submarines, heavy artillery, and tanks—

should be ceded to the League to form the *Head-quarter's Force,* and that no state should be allowed to own them or to make use of any new invention for warlike purposes.

"There should be no delay in handing over the new arms before they can claim long traditions. Vested interests have not yet been created on a permanent footing. Great disturbance would not be caused at present by the suggestion of denationalisation."

This really claims the advisability of verbal prohibition, which is absolutely useless, unless supported by the second class of safeguard, periodic "inspection." Major Davies suggests "all arsenals and munition factories would be open to inspection by the General Staff, who would use them, when necessary, for arming the quota of a nation other than that in whose territory they were situated." We know of no practical method by which inspection could be relied upon to give satisfactory warning of the conversion of the plants of the I.G. for war purposes. A distinction must be made between those weapons whose production can and cannot be practically controlled by inspection. In attempting such a classification, Major Davies claims, "It is difficult to prevent the secret manufacture of rifles, but it is easy to prevent the manufacture of tanks, aeroplanes, gas, or submarines." No one having witnessed the large scale operations of assembling tanks and heavy guns, and aware, at the same time, of the German methods of producing mustard gas or Blue Cross compounds, could make such an elementary mistake in classification, and any international disarmament arrangements based on such an error can only produce a false security. *Gas is the outstanding case of a weapon whose manufacture it is difficult to prevent.*

258

Chemical Warfare and Disarmament

"Vested Interests."—With regard to the vested interests in the new method of warfare, the most striking example is again the I.G. We find Ludendorff consulting Krupp and the I.G. representative when formulating his plans for a vast munition programme. Few people have realised the existence of another Krupp in the I.G. It would, indeed, be a revelation to find Germany sharing in these schemes of disarmament to the extent of voluntarily abandoning her dye monopoly. For such a situation is the only one consistent with safety. While the sole big source of production of these substances exists in Germany or in any one country for that matter, no scheme of disarmament is on sure ground.

"Handing Over" Inventions.—Certain disarmament advocates have ingenuous ideas with regard to new war inventions, and their "handing over" to the League. How can an invention be handed over? If every country informed the League of its new scientific war developments, those countries would still be aware of them. It is possible, commercially, to hand over any invention by assigning a patent, but this is of no use for war purposes. What country would regard patent law as a barrier to the use of a valuable war invention? Secondly, the cession of an invention to the League depends entirely on the goodwill of the nation concerned. No country can be sufficiently inspected to root out its new inventions. Suppose a gas ten times more useful, from a military point of view, than mustard gas were discovered in the laboratories of the I.G. An inspector, or "Secret Service" agent, at the next bench in the laboratory might never know that the research was not aimed at the discovery of a new dye. World equilibrium may at this moment be threatened by the discoveries of some absorbed

259

scientist working, say, in a greenhouse in St. John's Wood.

We come back to the same point, that the crux of the situation lies in the possession of the means of production. There is hope of controlling this for a weapon like a tank, but it cannot be controlled for chemical warfare. If the League requires these weapons it cannot rely on obtaining them from a monopoly source so complete as the I.G. Further, with or without a League the mere existence of this monopoly is a permanent menace to peace.

Neglect of Chemical Disarmament in the Treaty. —Let us face the facts. Our treatment of chemical industry during the Treaty negotiations and in the Treaty itself persistently ignored its chameleon nature. We knew that the nitrogen plants at Oppau and Merseburg were the most menacing munition plants in existence. We knew the grave dangers of leaving Germany, a guilty country, in possession of the poison gas monopoly. Yet, deaf to such arguments, the Treaty opportunity was ignored. Even now the lesson is only half learnt by those whom it vitally concerns.

Here is a new weapon whose exploitation demands research and large scale production. The former cannot be checked, and the latter cannot be destroyed or suitably controlled to prevent conversion for war purposes. Yet three distinct features of this weapon make the disarmament need imperative.

In the first place, everything points to "chemical disarmament" as a key measure to control the large scale use of all other weapons. The aggressive agent in war is the chemical. All weapons, except the bayonet, depend upon it.

In the second place, chemical warfare is itself so overwhelmingly important that it is farcical to contemplate any disarmament scheme which does not, first and foremost, tackle this question.

Thirdly, no nation ever held a more complete monopoly for any weapon than did Germany for chemical warfare. Yet the levelling up process which occurred during the war, tending towards armament equilibrium, towards removal of enormous disparity, failed to touch the chemical arm. Germany through her guilty exercise of the new weapon, has still further increased her enormous manufacturing superiority for war.

This age has witnessed the growth of an industry critical for war and disarmament. Others will follow as science progresses. Without them, the possibility of sudden decisions, and therefore war incentive will be removed. Sir Oliver Lodge prophesies the war use of the newly controlled atomic energy. The fulfilment depends on the growth of another critical war industry whose nature it would be difficult to foretell. It is these critical industries which rational disarmament must harness. At present the chemical industry holds the field.

Surely the first and crying need is to effect a redistribution of these organic chemical forces. This, indeed, is the one solid chemical disarmament measure which can and must be brought about.

The certain establishment of these industries in the chief countries outside Germany must be fixed far beyond the hazard of local politics and the reach of organised German attack. True, it is essential that no such support should in any way drug the will, weaken the initiative and impoverish the service of the fostered industries. This must depend upon wise organisation and control in the country concerned.

I claim, however, that it is one of the main duties of any League of Nations or other organisation deal-

261

ing with disarmament to proceed two steps beyond the paragraph in Article 8 of the Covenant. This runs as follows: "The members of the League undertake to interchange full and frank information as to the scale of their armaments, their military, naval, and air programmes, and the conditions of such of their industries as are adaptable to warlike purposes." Such an exchange of information must be used, first, to isolate that industry which is of a vital or key nature to the armament of the period, either on account of its value as a universal check, or because it fosters some particularly deadly new type of weapon or aggressive agent. The chemical industry at present fulfils both conditions, for without it, all weapons except the bayonet become silent, and it includes the organic chemical industry which fosters the deadly weapon of the period.

Secondly, rational disarmament must prevent the existence of monopoly in this critical industry. It may be objected that we are interfering with the play of ordinary economic laws. But we must face the possibility that the war of the future can never be averted without such interference. Indeed, if we accept the reports of the American Alien Property Custodian, this very monopoly which now threatens us was established by methods open to the same objections. It is indeed an interesting question whether the German dye monopoly resulted from forces which directly opposed the play of economic law. Further, the question is not so simple as it appears, for, in the industries which disarmament most concerns, governing technical changes are constantly occurring, and the normal home for the production of a whole range of chemical products may be shifted by a change of process which demands new raw materials or new types of energy and power. We must be ready, in certain critical cases, to regard disarmament as the paramount need. International agreement, through the League or otherwise, must find a suitable method to control the critical industry and prevent its use against world peace.

To be the ardent possessor of an ideal, to be its official guardian, does not allow us to ignore the technical aspect of an international and national issue. After our gigantic praiseworthy, but wasteful, attempts at chemical armament, let us at least disarm on rational lines.

CONCLUSION

THE TREATY AND THE FUTURE

I have endeavoured to present the facts of chemical warfare as briefly yet as truly as possible, giving a glimpse of the war possibilities inherent in this branch of applied chemical science. Nor have I ignored the hidden forces which inspired, stimulated, and supported the huge war chemical experiment. The great Rhine factories of the I.G. still cast their shadow on the outer world, obscuring the issues of reconstruction. This looming menace, its share in the past and future of chemical warfare, and the fatal growth of the latter present questions demanding an imperative answer. It is the weak point of world disarmament.

The Treaty of Versailles answers the riddle in principle, but have the actual clauses been unfulfilled?

Article 168 demands the limitation of munitions production to factories or works approved by the Allied and Associated Governments. "All other establishments for the manufacture of any war material whatever shall be closed down."

True, the plants of the I.G., like most other munition plants, have a dual function for peace and war. But their recent vital use for the latter brings them without doubt within the scope of the above clause. Are they still equipped for war purposes? Very drastic action will have been necessary by the Inter-Allied Commission of Control to justify a negative answer. Has that action been taken? If not, the I.G., a second Krupp, remains in splendid isolation,

264

secure behind our mediæval but generous conception of munitions, for fifty per cent. of the German shell fillings, the message of their guns, were eventually provided by the I.G. It is true that they were manufactured in synthetic dye and fertiliser plants, but the explosives were none the less violent and the poison gases none the less poisonous. Do we understand that the Allied and Associated Governments voluntarily leave Germany in unquestioned possession of this vast source of munitions in the face of the Treaty Article 168?

Article 169 wisely requires that any special plant intended for the manufacture of military material, except such as may be recognised as necessary for equipping the authorised strength of the German Army, must be "surrendered to be destroyed or rendered useless." The most formidable examples of such excess production were, and remain, the nitrogen fixation and the nitric acid plants of the I.G. The factories of the latter represent explosives and poison gas capacity far in excess of the authorised needs of the German Army. Why, then, should they be left intact?

What is the authorised equipment of the German Army? In the first place the manufacture and use of poison gas is specifically forbidden by the Treaty. The plants in question are therefore all in excess of authorised production, and should be destroyed or rendered useless. At present, to the best of our belief, they stand ready to produce at short notice at the rate of more than 3000 tons of poison gas per month. Does this mean that we admit them as authorised equipment? If so, we are ourselves contravening another clause of the Treaty.

The Treaty tabulates the authorised equipment in stock of shell. Based on the figures, we find that the actual war explosives production of the I.G., which, we believe, still largely remains available, could meet the total stock allowed to Germany by the current production of little more than one day!

Even if the Treaty provided authority, could these plants evade their just penalties on the ground of commercial world need?

Consider the question of German poison gas, all produced within the I.G., and its use and manufacture in Germany forbidden by the Treaty. It was made in converted or multiplied dye plants, or in special plants of the same type. Germany's great advantage was due, unquestionably, to her pre-war dye monopoly. The 1913 figures for production and home consumption are given below, under (A) and (B):

Country.	A Dye Production, 1913. Tons	B Home Dye Consumption. Tons	C Dye Production, 1918, Tons
Germany	135,000	20,000	135,000 (probably more)
Switzerland	10,000	3,000	12,000
France	7,500	9,430	18,000
U.K.	4,500	31,730	25,000
U.S.A.	3,000	26,020	27,000
Other Countries	3,000	72,820	4,000
Total	163,000	163,000	221,000

The completeness of the German monopoly stands clearly revealed. If, therefore, any plants capable of making dyes were built for poison gas or explosives during war, they could find no post-war *raison d'être* unless the feeble production of other countries had even further diminished.

Do the above figures (C) justify such an assumption? There is an increase of production outside Germany of nearly 60,000 tons per annum. Almost all of this, representing development under definitely expressed national policy, must be maintained unless we wish to revert to the exceedingly dangerous situation of a German dye and poison gas monopoly. Much of this 60,000 tons per annum German excess could be covered by plants used or built specially for poison gas or explosives.

There is every reason, for world peace, to eliminate such excess plants. There is no important reason, for commerce, to maintain them. In addition, many of them represent excess capacity which should be destroyed because they originated solely for the exploitation of a forbidden weapon. Even if a generous ruling, superimposed on the Treaty, offered these guilty plants a new lease of life because of their urgent peace-time use, the claim could not be supported before neutral experts. The Treaty provides authority for the disarming of certain chemical munition plants. Nothing but the most drastic economic need can justify departure from this critical disarmament measure. The need may justify Treaty exemption for other types of munition production in which the disarmament aspect is not so overwhelmingly important. The matter demands examination. We can hardly conceive that this has not been done. Are our missions equipped to meet the best German commercial minds on such a matter? In any case, Allied Governments have already wisely adopted a dye industry policy inconsistent with the special Treaty immunity of the excess I.G. munition plants. Our figures remove any ground for the economic argument.

The nitrogen fixation plants of the I.G. undoubtedly demand the same critical examination. These plants were built almost entirely for war purposes, for the production of ammonia to be oxidised to nitric acid. Ammonium nitrate also resulted. These sub-

267

stances are the mainstay of explosives warfare, and, as a matter of fact, their production in these very plants was the chief factor which enabled Germany to continue the war beyond 1915.

Under the simple reading of the Treaty clauses, the plants should "be destroyed or rendered useless." Here, possibly, strong arguments will be advanced by Germany for the retention of the plants for the purpose of fertilising her own soil. The argument is strong, for the impoverishment of German soil has been such as to demand, theoretically, enormous tonnages of ammonium sulphate. But it is vital, for the stability of peace, that this unique capacity for producing explosives must not remain the monopoly of any one country. It is the expressed intention of certain governments outside Germany to foster the nitrogen fixation enterprise. If, then, we admit the immunity of these German plants from the Treaty, for strong agricultural reasons, we must not allow Germany to use this privilege as a military advantage.

In other words, if we yield to such arguments it must be on two conditions. In the first place, the plants to evade the Treaty clauses must be proved necessary for German agriculture. Secondly, the products of the untouched plants must be used for this purpose and no other. As far as we know, no attempt has been made to apply the Treaty to the nitrogen fixation plants, and their products, instead of being mainly used for agriculture on German soil, have served as a deliberate weapon against the growing chemical industries of other countries.

Indeed, the figures at our disposal would indicate that even if the full demands of German agriculture were met, the plants built and projected leave a big margin which can only find outlet by export or military use. According to the *Frankfürter Zeitung* of

268

November 23rd, 1919, the total consumption of nitrogenous material by Germany was, in 1913, as follows:

Source and Nature	Tons	Tons Calculated as Nitrogen
Chili Saltpetre	750,000	116,000
Ammonium Sulphate	460,000	92,000
Norwegian Nitrate	35,000	4,500
Calcium Cyanamide	30,000	6,000
Haber Ammonium Sulphate (by Fixation)	20,000	4,000
Total		222,500

The same journal, October 18, 1919, states the capacity of the finished Haber plants to be equivalent to 300,000 tons of nitrogen per annum, and the total consumption of the old German Empire was thus less than the amount available from one source alone, *i.e.* nitrogen fixation by the Haber process. But other pre-war German sources of nitrogen, expanded by the war, will easily contribute their pre-war quota. We can therefore very safely assume German capacity of above 400,000 tons of nitrogen per annum, approximately twice the pre-war consumption. It is exceedingly unlikely that Germany will actually consume such a quantity. In any case, a large excess is now deliberately used to recapture world chemical markets, and this, as explained above, should be dealt with under the Treaty even if special immunity be afforded the capacity required for home purposes. We are indeed entitled to ask, what is being done on this vital matter?

Article 170 prohibits the importation of munitions of every kind into Germany. Considered from the point of view of chemical munitions, this clause shows a complete failure to understand the situation. Far from importing, possession of the I.G. leaves Germany the greatest potential exporter of chemical

269

munitions in the whole world. Further, it is not improbable that countries outside Germany may encourage her in munitions production for export. Lord Moulton stated in a speech at Manchester in December, 1914: "Supposing our War Minister had been in the last few years buying in the cheapest market for the sake of cheapness, and that he had had the munitions of war manufactured by Krupp's of Essen. Gentlemen, I think he would have been lynched about three months ago."

We have fallen far from the inspired resolution of those days! Knowing the true war significance of the I.G. as a second Krupp, if we fail to establish our own organic chemical industries, that warning may become a prophecy.

Article 171 forbids the manufacture of asphyxiating gases and analogous materials in Germany.

Has this clause any value unsupported by definite measures of control? With such an enormous capacity of rapidly convertible production, need Germany consider the production of these chemicals during peace? Once engaged in war, what is the value of the prohibition? True, failure would imply penalties for the specific breach of the Treaty. But a similar breach of International Convention is already involved, and admitted in the first phrase of Article 171: "The use of poison gases being prohibited, etc."

It is difficult to see, therefore, unless penalties be actually incurred for the existing breach, why Article 171 would be a serious deterrent for the future.

A trenchant comparison is afforded by the motive for this Treaty Article, and the actual operation of other Articles which should support it.

The Treaty makers thought it necessary to give direct reference to chemical warfare. They issued a special edict against its use. This alone should have guided those responsible for the execution of the Disarmament Clauses of the Treaty, measures of general application to the means of production of the different types of weapon. Have the special plants erected for poison gas received drastic action under the Treaty? It is to be feared that they and other war chemical plants of the I.G. have received undeserved immunity.

Where lies our help apart from the Treaty? World peace depends upon disarmament. True peace must come from a radical change in the outlook and sentiment of individuals. The forces working through these channels are the real peacemakers. But a League of Nations can forward the cause by wise measures of disarmament, and this implies limiting war producing capacity. The weak point in such a scheme is the organic chemical industry. There must be a redistribution of capacity, for while Germany retains a vast world monopoly of potential organic chemical munitions, which fed the armaments of the past with explosives and poison gas, and to which the weapons of the future are looking for inspiration and sustenance, disarmament will be a hollow farce.

The League of Nations may succeed in rooting out the means of production of certain munitions. But organic chemical factories must survive for the sake of their material contribution to the welfare of humanity. They cannot be inspected and controlled, as we have shown, and there is only one sound solution. The obstacle to peace must be removed by decentralising the organic chemical factories. We cannot leave this monopoly in the hands of any country. It now lies a weapon ready to the hands of those who created and wielded it with such success. Redistributed, this dangerous productive grouping will create a source of stability and strength to a League of Nations, and will invite a national sense of security, so essential to peace and disarmament under the present régime. This has only one meaning, the establishment of dye industries in Allied countries. This may clash with certain political schools of thought developed before the war without a due realisation of the organic way in which production links up with national defence. But let there be no misunderstanding. The refusal to support this critical industry is a definite sacrifice of vital national issues. Political principles responsible for such opposition no longer merit the name; they have become a fetish.

Our armies repelled the German chemical attack. They stood and fell unprotected before the early German clouds and unprotected again before the vile contact of mustard gas. The awful price they paid for our safety demands that we do more than rest contented with the sacrifice. It is an imperative and patriotic duty to the fallen, to the future of the race, and to the Empire, that, faced once again with modern war, we should be able to say, "every possible precaution was taken." But the chief precaution will have been neglected unless organic chemical industries are fostered on Imperial soil.

But what of chemical warfare itself? It is a growth, malignant or otherwise, according to our creeds, which will continue until very definite steps be taken to suppress it, with all war. Therefore, urgent guarantees for national safety are absolutely essential until the web of peace is strongly organised, which cannot be until the immediate menace of the monopoly in production is removed. But even then, until the general peace is fairly implanted, we must be ready for any surprise from an unscrupulous enemy. Research and training in chemical protection must be continued, and this can only be ensured by keeping abreast with offensive chemical warfare. "The Struggle for the Initiative" has at least established this.

Each nation and any League of Nations must seriously face the question of the establishment of elaborate and complex chemical warfare organisations. It seems to me that the logical course of thought and action is as follows. If guarantees are forthcoming, internationally, removing this grave German chemical warfare threat through her manufacturing monopoly, then the need for a definite chemical striking force and organisation will be greatly reduced. National safety is itself a corollary of world disarmament. But if satisfactory guarantees were forthcoming it would be consistent with national safety to limit the chemical warfare equipment of each nation to what would actually represent a scientific military brain. So long as national ministries for war or defence exist, they must possess even under the most stringent disarmament conditions, fully accredited within their regular staffs, an individual or individuals with scientific and military training, who represent knowledge, vision, and the power to expand in chemical warfare. What would be said of a great nation not equipped to think for the future on naval or artillery questions? Technical naval and military minds have evolved for these purposes. We are not slow to judge and act on the value of a new ship, tank, or machine-gun. The chemical arm is even more specialised and demands the same combination of scientific and military thinking and training. Whatever international disarmament decisions may be forthcoming, unless they seriously dismember the Defence Ministries, we should ensure that the pre-war position is corrected and that our staff conception and organisation covers the chemical weapon.

One alone of the Allied and Associated Powers was able to see the chemical menace with clear and unprejudiced vision. This was America, for she not only entered the war less hampered by traditions than the rest, but at a period when the chemical war was in full blast. More than a quarter of all her casualties were due to "gas," and no other arm produced as many in her ranks. As a result, we see America establishing an independent peace Chemical Warfare Service, as sister service to the Infantry and Artillery. This can only be interpreted as a frank realisation of the place of chemical warfare and of the need for serious international guarantees in the present situation.

Let us take a balanced view of the facts, realise the unique significance of chemical warfare and chemical industry, for war and disarmament, and act accordingly.

INDEX

Index

276

Index

Index

280

NAME INDEX

Name Index

Lambert, Major, 126.
Lebeau, Professor P., 101.
Levinstein, Dr. H., 168.
Livens, Major, 60.
Lodge, Sir Oliver, 94.
Ludendorff, General, 70, 82, 90, 91, 114, 147, 149, 259.

Meyer, Victor, 27
Macpherson, Captain, 121.
McConnel, Lieut., 208.
Moulton of Bank, Rt. Hon. Lord, 5, 169, 242, 243, 270.
Moureu, M. Charles, 100.

Norris, Colonel, 206, 208, 209.

Ozil, General, 100, 105.

Palmer, Mitchell, 19, 189.
Paterno, Senator, 101.
Penna, Colonel, 101.
Pick, Dr. H., 125, 129, 130, 131.
Pollard, Professor A. F., 112.
Pope, Sir William, 165, 191, 202.

Ramsay, Sir William, 94.
Rayleigh, Lord, 94.
Runciman, W., 146.

Sachur, Professor, 35.
Schmaus, Lieut. Dr., 75.
Schwarte, see Technik im Weltkriege (Subject Index).
Schweitzer, Dr. Hugo, 194, 195, 211.
Sering, Dr. Max, 211.
Stieglitz, Professor Julius, 191, 198, 200.

Thomas, Albert, 100.
Thorpe, Prof. J. F., 96, 99.
Thuillier, Major-General H. F., 94, 98, 105.

Villavecchia, Prof. 101.
Vincent, Monsieur, 100.

Watson, Colonel, 93.
Weiss, M., 100.
Wells, H. G., 112.
Wing, Major-General, 43.